Dynamic Population Models

# THE SPRINGER SERIES ON

# DEMOGRAPHIC METHODS AND POPULATION ANALYSIS

**Series Editor**

## KENNETH C. LAND

## Duke University

In recent decades, there has been a rapid development of demographic models and methods and an explosive growth in the range of applications of population analysis. This series seeks to provide a publication outlet both for high-quality textual and expository books on modern techniques of demographic analysis and for works that present exemplary applications of such techniques to various aspects of population analysis.

Topics appropriate for the series include:

- General demographic methods
- Techniques of standardization
- Life table models and methods
- Multistate and multiregional life tables, analyses and projections
- Demographic aspects of biostatistics and epidemiology
- Stable population theory and its extensions
- Methods of indirect estimation
- Stochastic population models
- Event history analysis, duration analysis, and hazard regression models
- Demographic projection methods and population forecasts
- Techniques of applied demographic analysis, regional and local population estimates and projections
- Methods of estimation and projection for business and health care applications
- Methods and estimates for unique populations such as schools and students

Volumes in the series are of interest to researchers, professionals, and students in demography, sociology, economics, statistics, geography and regional science, public health and health care management, epidemiology, biostatistics, actuarial science, business, and related fields.

The titles published in this series are listed at the end of this volume.

# Dynamic Population Models

*by*

Robert Schoen
Department of Sociology
The Pennsylvania State University
University Park, PA, USA

Springer

A C.I.P. Catalogue record for this book is available from the Library of Congress.

ISBN-10 1-4020-5229-4 (HB)
ISBN-13 978-1-4020-5229-3 (HB)
ISBN-10 1-4020-5230-8 (e-book)
ISBN-13 978-1-4020-5230-9 (e-book)

---

Published by Springer,
P.O. Box 17, 3300 AA Dordrecht, The Netherlands.

*www.springer.com*

*Printed on acid-free paper*

CONTENTS

Preface ........ xi

List of Tables and Figures ........ xiii

**1. Population Models With Constant Rates ........ 1**
1.1. Introduction ........ 1
1.2. Some Basic Concepts ........ 1
1.3. The Basic Life Table ........ 3
1.4. Multiple Decrement and Cause Eliminated Life Tables ........ 7
1.5. Stable Population Models ........ 9
1.5.1. Continuous Stable Models ........ 9
1.5.2. Discrete Stable Models ........ 12
1.6. Multistate Life Tables ........ 14
1.6.1. Scalar Multistate Equations ........ 14
1.6.2. Matrix Multistate Equations ........ 16
1.6.3. Multistate Expectancies and Summary Measures ........ 19
1.7. The Multistate Stable Population ........ 20
1.8. Summary ........ 22
1.9. Exercises ........ 23

**2. Convergence to Stability ........ 27**
2.1. Introduction ........ 27
2.2. Conditions for Stability ........ 27
2.3. The Speed of Convergence ........ 29
2.4. Some Explicit Analyses of Convergence ........ 33
2.4.1. Convergence in 2 × 2 Leslie Models ........ 34
2.4.2. Convergence in a 3 × 3 Leslie Model ........ 34
2.4.3. Convergence in a Two Living State Multistate Model ........ 37
2.5. Summary ........ 40
2.6. Exercises ........ 41

**3. Population Momentum ........ 43**
3.1. Introduction ........ 43
3.2. Momentum As a Demographic Phenomenon ........ 43
3.2.1. The Basic Dynamics ........ 43
3.2.2. Momentum and Aging ........ 46

3.3. Momentum Under Gradual Declines to Stationarity .................... 50
3.4. Spatial Momentum .................... 51
3.4.1. Spatial Momentum in Two Region Models Without Age .................... 52
3.4.2. Decomposing Total Multistate Population Momentum .................... 53
3.5. Summary .................... 57
3.6. Exercises .................... 57

**4. Demographic Change at the Margin .................... 59**
4.1. Introduction .................... 59
4.2. Marginal Change in Fixed Rate Models .................... 59
4.2.1. Changes in Stable Population Measures .................... 60
4.2.2. Changes in Stable Population Characteristics .................... 62
4.3. Recognizing Age-Dependent Growth .................... 63
4.4. Connections Between Populations Having the Same Vital Rates .................... 65
4.5. Marginal Change in Dynamic Populations .................... 67
4.5.1. Derivatives of Aggregate Demographic Measures .................... 67
4.5.2. Marginal Change and Stabilization .................... 70
4.6. A Spectral Decomposition of Marginal Change .................... 72
4.7. Summary .................... 74
4.8. Exercises .................... 75

**5. Longevity and Dynamic Mortality .................... 79**
5.1. Introduction .................... 79
5.2. An Overview of Changes in Human Mortality .................... 79
5.3. A Third Analytical Perspective: The Wedge-Period .................... 82
5.4. Period-Cohort Considerations .................... 82
5.4.1. The Period-Cohort Contrast in Mortality .................... 82
5.4.2. Average Cohort Life Expectancy .................... 83
5.5. A Shifting Gompertz Mortality Change Model .................... 85
5.5.1. The Bongaarts-Feeney Approach .................... 85
5.5.2. Relationships in a Shifting Gompertz Model .................... 89
5.6. Models of Constant Mortality Decline Over Time .................... 92
5.6.1. Three Models With Linearly Changing Mortality .................... 92
5.6.2. Gaps and Lags Between Period and Cohort Life Expectancies .................... 94
5.6.3. The Modal Age At Death .................... 96
5.7. Summary .................... 98
5.8. Exercises .................... 99

**6. Timing Effects on Fertility, Marriage, and Divorce.............. 101**
6.1. Introduction ........ 101
6.2. The Period-Cohort Contrast ........ 101
6.3. The Bongaarts-Feeney Approach ........ 103
6.4. Average Cohort Fertility ........ 105
6.5. Comparing the Bongaarts-Feeney and ACF Approaches ........ 106
6.5.1. Timing Effects in Selected Models ........ 107
6.5.2. Timing Effects in the United States, 1917–97 ........ 107
6.6. Examining Parity-Specific Period TFRs ........ 110
6.7. A Diminishing Population Whose Every Cohort More Than Replaces Itself ........ 112
6.8. Timing Effects on First Marriage ........ 113
6.8.1. Measuring Timing Effects on First Marriage ........ 114
6.8.2. Timing Effects on First Marriage in England and Wales and the United States ........ 115
6.9. Timing Effects on Divorce ........ 116
6.9.1. Measuring Timing Effects on Divorce ........ 118
6.9.2. Timing Effects on Divorce in the United States ........ 119
6.10. Summary ........ 120
6.11. Exercises ........ 121

**7. Dynamic Birth-Death (Multi-Age) Models ...................... 123**
7.1. Introduction ........ 123
7.2. Models Where the Age-Specific Rates Are Implicit ........ 123
7.2.1. Considering Models With a Constant Generation Length ........ 124
7.2.2. Cyclical Models With a Constant Generation Length ........ 126
7.2.3. Intergenerational Relationships in a Simple Model With Sinusoidal Births ........ 128
7.3. Metastability, A Closed Form Extension of Stability ........ 132
7.3.1. Deriving the Discrete Metastable Model ........ 132
7.3.2. The Continuous Metastable Model ........ 135
7.3.3. Modeling Fertility Transitions and Their Associated Momentum ........ 138
7.4. Intrinsically Dynamic Models, An Alternative Extension of Stability ........ 142
7.4.1. The Two Age Group Intrinsically Dynamic Model ........ 143
7.4.2. The Three Age Group Intrinsically Dynamic Model ........ 146
7.4.3. The n-Age Group Intrinsically Dynamic Model ........ 148
7.4.4. IDM Birth Trajectories ........ 149

7.4.5. Observed Populations as IDM Populations ..................... 150
7.4.6. Transitions and Momentum in IDM Populations............ 150
7.5. Finding Rates Consistent With a Known Birth Trajectory .......... 152
7.5.1. Hyperstable Models ........................................................ 153
7.5.2. Models With Proportional Rates ...................................... 155
7.6. Summary ........................................................................ 155
7.7. Exercises ........................................................................ 156

**8. Dynamic Multistate Models .................................. 161**
8.1. Introduction ........................................................................ 161
8.2. Cohort Multistate Models With Proportional Transfer Rates ....... 161
8.2.1. Relationships in Proportional Transfer Rate (PTR) Models 162
8.2.2. PTR Models With 2 Living States .................................. 165
8.2.3. Hierarchical PTR Models ................................................ 166
8.2.4. A PTR Model of Marriage, Divorce, and Remarriage....... 170
8.3. Multistate Metastable Models ................................................ 172
8.4. Multistate Intrinsically Dynamic Models.................................. 174
8.4.1. The Restricted Two Living State Uniform Growth Model 174
8.4.2. Additional Two Living State Models................................ 178
8.4.3. The Restricted No Growth IDM With Three Living States 180
8.4.4. The Restricted Uniform Growth IDM With n Living States ........................................................................ 181
8.5. Multistate Hyperstability ........................................................ 181
8.5.1. Deriving the Multistate Hyperstable Equations ................. 181
8.5.2. Relationships Between the Multi-Age and Multistate Hyperstable Models ........................................................ 184
8.5.3. A Further Examination of the Dynamic Two Living State Model ........................................................................ 185
8.6. Finding Rates Consistent With Multistate Population Distributions ........................................................................ 188
8.6.1. Estimating Multistate Arrays Through Iterative Proportional Fitting ........................................................ 189
8.6.2. Estimating Multistate Rates By the Relative State Attraction (RSA) Method ................................................ 190
8.6.3. Evaluating the IPF and RSA Approaches.......................... 193
8.7. Summary ........................................................................ 196
8.8. Exercises ........................................................................ 197

**9. Dynamic Models With Multiple Ages and States ................ 203**
9.1. Introduction ........................................................................ 203
9.2. The Multi-Age and State Hyperstable (MASH) Model ............... 204

9.2.1. Specifying the MASH Model .......... 204
9.2.2. An Illustrative Application .......... 207
9.3. The Multi-Age and State Metastable (MASM) Model .......... 209
9.4. Summary and Overview .......... 212
9.4.1. Chapter Summary .......... 212
9.4.2. Overview and Recapitulation .......... 212
9.5. Exercises .......... 214

**Appendix A Matrices and Their Eigenstructure .......... 217**

**Appendix B Selected Answers to Exercises .......... 221**

**Appendix C Index of the Principal Symbols Used .......... 225**

**References .......... 231**

**Author Index .......... 241**

**Subject Index .......... 245**

# PREFACE

Dynamic population modeling, the study of populations with changing vital rates, is an emerging area of research in mathematical demography that presently lacks an integrative, book length treatment. This volume seeks to fill that gap. It can be seen as a continuation of my earlier monograph, *Modeling Multigroup Populations* (Schoen 1988a), which focused on models with constant rates. The chapters in this book, after summarizing the essential aspects of fixed rate models, describe expanding efforts at introducing behavioral change. The discussion proceeds from studies of one-time changes at the margin and population momentum, through analyses of the effects of cohort timing on period measures, to the development of fully dynamic models with multiple ages and states. A demographic perspective is adopted throughout, with an emphasis on human populations at the aggregate level, exogenous rates of behavior, and deterministic models. Use is made of both continuous and discrete formulations. Only one-sex models are covered, however, as two-sex population models were extensively treated in Schoen (1988a); readers interested in more recent developments are referred to Iannelli, Martcheva, and Milner (2005).

This book is intended to be as accessible as possible. It assumes a very basic knowledge of calculus, but readers can find those essentials in Appendix A of Schoen (1988a). Appendix A of this book provides the necessary background in matrix algebra, an area unfamiliar to many demographers but necessary for discussions of dynamic models. To reinforce concepts introduced in the text, or to show additional relationships of interest, exercises are included at the end of every chapter. Selected answers are provided in Appendix B. Notation remains an issue, as there is no standard way to represent many of the terms and concepts discussed. The book tries to be reasonably consistent, but is not always successful. To help the reader follow the symbols employed, Appendix C provides brief definitions and references to the equation in which the symbol was introduced.

My work in mathematical demography has been advanced by many colleagues, students, and institutions. My greatest debt is to the late Young J. Kim, an outstanding mathematical demographer and biostatistician. I was fortunate to be able to collaborate very closely with her for ten years, and her contributions are manifest throughout this volume. Vladimir Canudas-Romo worked closely with me for two years while on a post-doctoral fellowship, provided innovative ideas and considerable energy, and furnished very helpful comments on Chapter 5. Stefan H. Jonsson was an indefatigable and extraordinarily resourceful collaborator and Graduate Research Assistant. Claudia Nau was very helpful in verifying a number of relationships and assembling the materials used in the book. I would also like to express my appreciation to Juha M. Alho, Cristina Bradatan, Alice Yen-hsin Cheng, Rachel Durham, Juan Pablo Lewinger, Kenneth C. Land, Xianbin Li, Hui-Peng Liew, P. Sankara Sarma, Kenneth D. Smith, Nicola Standish, Zenas Sykes, Paula Tufis, Robin Weinick, Howard Weiss, and Yan Yan. Johns Hopkins University, through its Department of Population Dynamics and Hopkins Population Center, provided an excellent research environment for me during the years 1989–1999. Since 1999, I have been at The Pennsylvania State University, and have benefitted greatly from the outstanding support and facilities provided by its Department of Sociology and Population Research Institute, and for the Sabbatical leave granted to me during Spring 2006. I especially want to thank Arnold S. and Bette G. Hoffman for their generosity in establishing the Hoffman Professorship that I currently hold, whose support has greatly facilitated my work. I am indebted to the National Institute of Child Health and Human Development (NIH) for its sustained support through grants HD19145 and HD28443. Evelien Bakker of Springer provided helpful advice, a positive attitude, and unfailing good cheer. My ultimate thanks are to my wife, Delores C. Schoen, for her encouragement and unwavering support.

Robert Schoen, PhD, ASA
Hoffman Professor of Family
Sociology and Demography
The Pennsylvania State University
University Park PA (USA)
May 2006

# LIST OF TABLES AND FIGURES


| | | |
|---|---|---|
| Table 2.1. | Demographic Measures Related to Convergence for 3 Selected Female Populations | 33 |
| Figure 2.1. | Birth Trajectories to Stationarity in the Neighborhood of a Double Root, Leslie Model with 3 Age Groups | 36 |
| Table 3.1. | Momentum and Ultimate Proportion Urban Values for Hypothetical Populations By Initial Percent Urban, Urban/Rural Fertility Differential, and Level of Urban/Rural Migration | 56 |
| Figure 3.1. | Cumulative Initial and Ultimate Populations, By Age, Based on Rates Observed in Singapore 1957 | 47 |
| Figure 3.2. | Projection to Stationarity of the Observed Population of Singapore 1957 | 49 |
| Table 4.1. | Time Derivatives of Population Functions | 70 |
| Table 5.1. | Summary Measures of Mortality for Life Expectancies of 20 to 80 Years, Coale-Demeny Model "West" Female Life Tables | 81 |
| Table 5.2. | Functions and Relationships in a Shifting Gompertz Mortality Change Model | 91 |
| Figure 5.1. | Observed and Coale-Demeny Model "West" Life Table Age-Specific Probabilities of Death in a High and a Low Mortality Population | 80 |
| Figure 5.2. | Cohort and Period Life Expectancy, ACLE, and CAL for a Sinusoidally Declining Mortality Model with Cohort Cyclical Fluctuations of 60 Years and Parameters $A = -11.215$, $b = 0.1$, $c = 0.008$ and $d = 0.25$ | 84 |
| Figure 5.3. | Stylized $\mu$ and $\nu$ Curves and the Distances Between Them, Shifting Gompertz Mortality Change Model | 86 |
| Figure 5.4. | Gaps and Lags Between Period and Cohort Life Expectancies | 95 |
| Figure 5.5. | Life Table Survivorship Functions and Distributions of Deaths for a Linearly Declining Gompertz Mortality Model with Parameters $A = -11$, $b = 0.1$ and $c = 0.01$ | 97 |
| Figure 6.1. | Values of the Mean Age at Childbearing (MAC), Bongaarts-Feeney Adjusted Fertility (TFR*), and Average Cohort Fertility (ACF) in a Model Population with a | |

Constant Period TFR of 1 that Experiences Cyclical Changes in the Timing of Period Fertility Beginning at Time 50 108
Figure 6.2. Values of the Period Total Fertility Rate (PTFR), the Average Cohort Fertility (ACF), the Cohort Total Fertility Rate for the Cohort Born in the Year t-26, and the Bongaarts-Feeney Adjusted Measure (TFR*), United States, 1917–2001 109
Figure 6.3. Annual Number of Births in Linearly Shifting Fertility Models by NRR and Annual Change in the Mean Age of Childbearing 113
Figure 6.4. Period and Cohort Proportion Ever Marrying (PEM) and Timing Adjusted PEM* Trajectories 117
Figure 6.5. Period and Cohort Probability of a Marriage Ending in Divorce (PMED) and Adjusted PMED* for Women Under Two Future Scenarios, Linear Increases and Decreases of 20% in Divorce Rates from 2000 to 2020, United States 1910–2000. 119
Table 7.1. Maximum Dollar Gain and Loss Values (Benefits Minus Contributions) Under Pay $45 and Get $45 Plans, When Amplitude (b) is 0.3, By Cycle Length (T) in a Cyclically Stationary Population 131
Table 7.2. Measures of Population Momentum in Hypothetical Populations Experiencing Exponential Fertility Declines Over Age and Time That Require Y Years to Achieve Replacement Level Fertility 141
Table 7.3. Age-Specific Fertility Patterns and their Variation With the Dominant Root in Three Age Group Intrinsically Dynamic Models 147
Figure 7.1. Exponentiated Sinusoidal NRR Trajectory, R(t), and Associated Birth Trajectory, g(t), for Generation Length $A = 30$, Amplitude $b = 0.2$, and Selected Cycle Lengths, T 126
Figure 7.2. Maximum Dependency Ratio (D) When Amplitude (b) is 0.3, for Values of Cycle Length (T) from 15 to 105 Years in a Cyclical Stationary Population Model 130
Figure 7.3. The IDM Birth Trajectory Following a Transition, at Time 0, From a Stable ($\lambda_A = 1.5$) to a Stationary ($\lambda_B = 1$) Regime, Under Fertility Patterns 2 and 4 151
Figure 7.4. The IDM Cumulative Mean Age of Childbearing ($\mu_{Mt}$) Following a Transition, at Time 0, From a Stable ($\lambda_A = 1.5$) to a Stationary ($\lambda_B = 1$) Regime, Under Fertility Patterns 2 and 4 152

Table 8.1. Selected Summary Measures from Actual 1988, Actual 1995, and Estimated 1995 Female Marital Status Life Tables for the United States 195
Figure 8.1. Life Course Schematic of Parity Distributions for Z Values From 0 to 10, United States Females, 1997 165
Figure 8.2. Proportion at Parities 0, 1, 2 and 3 Under Proportionalized Rates for United States Females, 1997 With Alternative Base Rates of Movement From Parity 0 to 1 169
Figure 8.3. Observed (obs) and Stylized (sty) Rates of First Marriage, Divorce, and Remarriage, with the Stylized Rates Based on Ages 34 to 38, United States Females, 1995 172
Figure 8.4. Life Course Schematic Showing Proportions Never Married, Married, and Divorced for Z Values From 0 to 6, United States Females, 1995 173
Figure 8.5. Time Trajectories of the Normalized Rate of Transfer from State 2 to State 1 ($m_{21}/M$) and the Fraction of Persons in State 1 ($\ell_1$), Sinusoidal Two Living State, No Growth Intrinsically Dynamic Model 177
Figure 8.6. Relative State Composition ($w_1$) and Total Population Size (N) for a Hierarchical Two Living State Proportional Eigenvector Model 187
Figure 8.7. Actual, Estimated, and Standard Marriage Rates From a Two-Living State (Married/Unmarried) Model Based on Rates for Swedish Women, Cohort Born 1930–1934 194
Figure 9.1. Summary Measures of the Trajectory of a Robust/Frail MASH Model 208

# CHAPTER 1

# POPULATION MODELS WITH CONSTANT RATES

## 1.1 INTRODUCTION

Population models are based on rates of birth, death, and movement, both movement from place to place and from status to status. Models can show how those basic demographic processes shape the size and composition of a population, revealing the underlying dynamics of growth and attrition, concentration and dispersal, aging and renewal.

Demography, the area of social science that focuses on populations, dates back nearly three and a half centuries to the *Natural and Political Observations* of John Graunt (1662). From Graunt's early life table, demographic models have grown in sophistication and in their ability to capture population behavior. Recently, demographers have begun to study dynamic models, i.e. models with behavioral rates that change over time. This volume traces a path that begins with fixed rate life tables and their extensions to stable and multistate models, passes through convergence to stability and the nature of demographic change at the margin, and leads to a range of dynamic models and the opportunities and challenges they present.

## 1.2 SOME BASIC CONCEPTS

The basic measures in demography are behavioral rates, e.g. birth rates, death rates, and marriage rates. Unless explicitly stated otherwise, the term rates in this book refers to occurrence/exposure rates, also known as central rates and rates of the first kind. Specifically, if D refers to the frequency of some event (e.g. births or deaths) and P represents the size of the population exposed to the risk of that event, then the rate, M, is given by

$$M = D/P \qquad (1.1)$$

Rates can, and usually are, made specific to population subgroups (Preston, Heuveline and Guillot 2001; Schoen, 1988a). For example, the death rate for men age 20–24 can be written

$$M_m(20, 5) = D_m(20, 5)/P_m(20, 5) \tag{1.2}$$

where the 20–24 year interval is understood to begin at exact age 20 and extend to the instant before the attainment of exact age 25, expressions of the form (x,n) indicate the interval beginning at exact age x and extending for n years, and the subscript *m* refers to males. For one-year age intervals, the width is usually not shown (e.g. $M_m(30)$ is the male death rate at age 30).

More fundamental and precise than rates, but not directly observable, are demographic forces, also known as intensities, instantaneous rates, or instantaneous probabilities. For example, the force of mortality at exact age x in a basic, decrement-only life table, denoted $\mu(x)$, is defined by

$$\mu(x) = [-1/\ell(x)]\,[d\ell(x)/dx] \tag{1.3}$$

where $\ell(x)$ represents the number of persons who survive to exact age x and d/dx is the conventional expression indicating differentiation with respect to x. The force of mortality thus represents the instantaneous risk of death at exact age x, expressed as the marginal proportional decline in the number who survive to exact age x. The minus sign is used so that the force is a positive quantity. The force of mortality at the midpoint of an interval approximates the rate over that interval, i.e.

$$\mu(x + n/2) \approx M(x, n) \tag{1.4}$$

Equation (1.3) is one form of the classic differential equation of demography, whose analysis is at the core of this book. When $\mu(x)$ and $\ell(x)$ are scalar functions, the solution is straightforward. Since $[-1/\ell(x)]\,[d\,\ell(x)/dx]$ is the same as $[-d\ \ln\ \ell(x)/dx]$, where ln denotes the natural logarithm, we can integrate equation (1.3) from x to x + n, exponentiate, and write

$$\ell(x + n) = \ell(x)\exp[-\int_0^n \mu(x + a)da] \tag{1.5}$$

The number of survivors to exact age x + n is equal to the number of survivors to exact age x times the negative exponential of the sum of the forces of mortality between ages x and x + n. Equation (1.5) expresses a general relationship that applies over intervals of any length. If one can assume that the force of mortality is constant between ages x and x + n, then the force of mortality can be approximated by M(x,n), and we have the constant forces solution

$$\ell(x+n) = \ell(x)\exp[-nM(x,n)] \qquad (1.6)$$

For the most part, demographic analyses have adopted one of two perspectives: cohort or period. A cohort is a closed group of persons who share some initial characteristic and are followed over time. [Unless explicitly noted otherwise, all of the models discussed are assumed closed to migration.] For example, a birth cohort may be composed of persons born in the same year, and a marriage cohort can consist of those married during a specified time interval. A period is a relatively short interval of calendar time, generally one year or several consecutive years. Marriages occurring in a given year, or death rates based on a three-year interval around a census, are examples of period data. In any one-year period, a number of cohorts are active. The interplay of period and cohort plays an important role in many social science analyses (cf. Ryder, 1965), and is the subject of Chapters 5 and 6. We now turn to the life table, which simultaneously embodies both the period and cohort perspectives.

## 1.3 THE BASIC LIFE TABLE

From the cohort perspective, a basic decrement-only life table follows a cohort of $\ell(0)$ persons from birth to the death of its last survivor, as the cohort is subject to attrition from a set of age-specific death rates. It is widely used in mortality analyses, and has been applied to many other subjects as well, including marriage and divorce, contraceptive use, educational attainment, and voting behavior. The starting value, $\ell(0)$, is known as the radix, and is usually chosen to be a large round number (often 100,000). Complete life tables show the number of survivors to every year. The more common abridged life tables typically show ages 0, 1–4, 5–9,. . ., 80–85, and 85 and over. With improved survivorship, the open ended interval increasingly begins at age 90.

The principal function of the life table is the survivorship function, $\ell(x)$, which gives the number surviving to exact age x. The remaining functions follow from the $\ell(x)$. The number of cohort deaths between the ages of x and x + n, d(x,n), is

$$d(x,n) = \ell(x) - \ell(x+n) \qquad (1.7)$$

The probability a person exact age x dies between the ages of x and x + n, q(x,n), is

$$q(x,n) = d(x,n)/\ell(x) \qquad (1.8)$$

The number of person-years lived by the cohort between the ages of x and x+n, L(x,n), reflects the cohort's exposure to risk, as a person-year is one year lived by one person. Stated precisely,

$$L(x,n) = \int_0^n \ell(x+u)du \qquad (1.9)$$

The life table death rate between the ages of x and x+n, m(x,n), is then

$$m(x,n) = d(x, n)/L(x, n) \qquad (1.10)$$

It is useful to define the function T(x) as the total number of person-years lived by the cohort at and above exact age x. Thus T(x) is the sum of L functions from age x to the end of the table, or the integral of $\ell(x)$ from age x through the highest attained age. The average life expectancy of a person exact age x, e(x), is then given by

$$e(x) = T(x)/\ell(x) \qquad (1.11)$$

To construct a life table from a set of age-specific death rates, it is necessary to transform M rates to q probabilities or, equivalently, express $\ell$ survivors in terms of L person-years. The procedure begins at age 0 and proceeds age by age. One could find L values by integrating the $\ell$ function, but that is generally not feasible unless the force of mortality is constant within age intervals. The General Algorithm (Schoen and Land, 1979), provides a framework for the $\ell$ to L transformation. In life table construction, it consists of *flow equations, orientation equations*, and *person-year equations*.

The flow equation of a decrement-only life table is simply equation (1.7), which equates $\ell(x+n)$ to $\ell(x)$ minus the flow to death in the interval. The orientation equation typically is

$$m(x,n) = M(x, n) \qquad (1.12)$$

which specifies that the life table reproduces the rates observed in the population at every age. Alternative assumptions are possible and have been used (e.g. Preston, Keyfitz, and Schoen 1972), but here we focus on the straight-forward relationship in equation (1.12).

The person-year equation represents the chosen solution to integral equation (1.9). Many choices are possible, but we emphasize three approaches (cf. Schoen 1988a). First, under constant forces, equation (1.6) applies and equation (1.9) can be integrated, yielding

$$L(x,n) = \ell(x)\{1 - \exp[-nM(x, n)]\}/M(x, n) \qquad (1.13)$$

The assumption that forces are constant within intervals can be rather strong for intervals of more than a year, and the constant forces assumption is generally considered a bit too crude for abridged (i.e. 5-year age interval) life tables.

A second approach is to assume that $\ell(x)$ is a linear function between ages x and $x+n$. Under linearity, equation (1.9) becomes

$$L(x,n) = (n/2)[\ell(x) + \ell(x+n)] \tag{1.14}$$

and the survivorship function is generated by

$$\ell(x+n) = \ell(x)\{1-(n/2)M(x, n)\}/\{1+(n/2)M(x, n)\} \tag{1.15}$$

A linear $\ell(x)$ function implies a uniform distribution of deaths within each interval, and hence an increasing force of mortality. Because mortality generally does increase with age above ages 10–12, the linear assumption has been found to produce acceptable abridged life tables. However, when forces of decrement are decreasing with age and intervals are more than a year in length, the linear method may not be appropriate.

The third alternative is the mean duration at transfer approach (cf. Schoen 1988a). Chiang (1968) proposed the useful function a(x,n), defined as the average number of years lived in the x to $x+n$ age interval by persons dying in that interval. Since the total number of person-years lived is the sum of those lived by those who die and by those who survive, we can write

$$L(x,n) = n\ell(x+n) + a(x, n)\ d(x, n) \tag{1.16}$$

Using equation (1.16), we can use Chiang's a to generalize the linear relationship in equation (1.15) to

$$\ell(x+n) = \ell(x)\{1-a(x, n)M(x, n)\}/\{1+[n-a(x, n)]M(x, n)\} \tag{1.17}$$

Above age 5, a(x,n) is typically close to n/2. However, the size of a(x,n) is influenced by the nature of the force of decrement in the interval. For a given level of M(x,n), a more steeply rising force of decrement produces a larger a(x,n). The Mean Duration at Transfer (MDAT) approach exploits that relationship and uses the values of the rates in adjacent intervals to estimate a(x,n). Specifically (Schoen, 1978)

$$a(x,n) = \{u(x, n)\ell(x) + w(x, n)\ell(x+n)\}/d(x, n) \tag{1.18}$$

where

$$u(x,n) = (n^2/240)\{M(x+n,n) + 38\ M(x,n) + M(x-n,n)\}$$

and $$w(x,n) = (n^2/240)\{14\ M(x+n,n) + 72\ M(x,n) - 6\ M(x-n,n)\}$$

Because it allows the force of decrement to increase or decrease and makes fuller use of the available data, the Mean Duration at Transfer method may yield better values in 5-year age intervals. Alternatively, usable estimates of Chiang's a can be found in a number of ways (cf. Exercise 15 in Section 1.9 and the discussion in Preston, Heuveline and Guillot (2001, Ch. 3). Judgment is always involved, and no method of life table construction should be used mindlessly. In particular, if rates are large, intervals of one year (or less) should be used. Indirect standardization (cf. Preston et al, 2001; Ch 2) is the simplest approach to estimating rates for shorter age intervals. More advanced techniques for graduating rates can be found in the classic actuarial text by Miller (1946).

Special procedures are needed at the youngest and oldest ages. As most deaths in infancy occur soon after birth, the linear assumption is not appropriate. Keyfitz (1970) suggested the empirical relationships

$$a(0) = 0.07 + 1.7\ M(0)$$
$$a(1,4) = 1.5 \qquad (1.19)$$

and alternative procedures are given in Preston et al (2001, Ch 3). To end the life table, a highest age is selected, typically 85 or 90. Since everyone dies, the number surviving to that highest age, say 90, is the number that will die above that age. Thus we can rewrite equation (1.10) as

$$L(90,\infty) = \ell(90)/m(90,\infty) \qquad (1.20)$$

to find the number of person-years lived in the open-ended age interval.

The life table can also be viewed from a period perspective. Consider a population that has a long history of constant births and age-specific mortality that does not change over time. Eventually that population becomes stationary, that is it attains a constant size and a fixed age composition. The number of persons in the stationary population is T(0), there are $\ell(0)$ births each year, and there are L(x,n) persons between the ages of x and $x+n$. The period interpretation of the L(x,n) function reinforces the rationale for its use as the denominator for age-specific rates. Each year, the stationary population depicts what every cohort experiences over its lifetime. For example, every year there are d(x,n) deaths between the ages of x and x+n, giving a total of $\ell(0)$ deaths. The stationary population (crude) rate of birth and death is then

$$M = \ell(0)/T(0) = 1/e(0) \quad (1.21)$$

Regardless of the age pattern of mortality, if e(0) = 80, then the overall stationary population death rate is 1/80.

As a stationary population, the life table goes beyond providing probabilities of death and expectations of life. It is a model for processes that can be assumed to change little from year to year. Hence it can be applied to such topics as the number of students in an educational system or the number of persons on a company's pension rolls.

## 1.4 MULTIPLE DECREMENT AND CAUSE ELIMINATED LIFE TABLES

Multiple decrement life tables allow persons to leave the table for different causes, e.g. for specified causes of death or, in the case of a nuptiality-mortality life table, for either first marriage or death. To recognize multiple causes of decrement, we first define the decrement rate specific to cause i, $m^i(x, n)$, as

$$m^i(x, n) = d^i(x, n)/L(x, n) \quad (1.22)$$

where $d^i(x, n)$ is the number of life table decrements to cause i between the ages of x and x+n. The denominator must be L(x,n) as all persons are at risk of decrementing from cause i. Causes must be exhaustive and mutually exclusive, so that

$$d(x,n) = \sum_i d^i(x, n) \quad (1.23)$$

To construct a multiple decrement life table, one needs observed age-cause specific rates $M^i(x, n) = D^i(x, n)/P(x, n)$, where $D^i(x, n)$ is the number of observed decrements to cause i between the ages of x and x+n, and we assume that $M^i(x, n) = m^i(x, n)$. After constructing a basic life table for the population, one can allocate the d(x,n) at each age to the different causes of decrement by

$$d^i(x, n) = d(x, n)[M^i(x, n)/M(x, n)] \quad (1.24)$$

The total number of decrements to cause i at and above age x is denoted by $\ell^i(x)$, and is given by

$$\ell^i(x) = \Sigma_j \, d^i(j) \quad (1.25)$$

where the sum over j goes from age x through the highest age in the table. The age-cause-specific force of mortality is defined by

$$\mu^i(x) = [-1/\ell(x)][d\ \ell^i(x)/dx] \qquad (1.26)$$

As is true for the decrements, the $\ell^i(x)$, $\mu^i(x)$, and $M^i(x, n)$ for every cause sum to the corresponding basic life table value.

Multiple decrement life tables provide a number of useful summary measures. The most commonly used is $\ell^i(0)$, the lifetime number of decrements due to cause i. The analogous person-year function, $L^i(x, n)$, is generally neither calculated nor used.

Questions are frequently asked about the effects of eliminating a cause of decrement. For example, how much would life expectancy increase if deaths from cancer were eliminated, or what proportion would ever marry if there were no mortality before age 50. Cause eliminated, or Associated Single Decrement (ASDT) life tables seek to answer such questions. Calculating such tables involves two distinct assumptions. The first is the vitality assumption. Nearly all ASDTs are calculated assuming that the force of decrement from the cause active in the ASDT is the same as the force from that cause in the multiple decrement life table. The force of decrement from the eliminated cause is set to zero. That assumption of unchanging vitality is nearly always counterfactual. For example, a cure for cancer would likely impact other causes of death, and the chance of ever marrying is known to be related to health. The analyst needs to chose the cause(s) to be eliminated so that the vitality assumption is defensible.

The second assumption underlying ASDTs deals with the additional exposure to the risk of decrement from the active cause(s). A number of techniques are available to accommodate the additional risk (cf. Schoen 1988a; Preston et al 2001). Because the approximate relationship in equation (1.4) applies to cause-specific forces of decrement, the simplest approach is to assume that the ASDT rate for active cause i is simply $M^i(x, n)$, the multiple decrement rate for that cause. The cause eliminated table can then be constructed using the $M^i(x, n)$.

The most commonly used summary measure provided by ASDTs is the increase in life expectancy resulting from the elimination of a cause. For example, Preston et al (1972, 770–71) show that e(0) was 73.78 years for U.S. Females under the death rates observed in 1964. Neoplasms (largely cancers) were the second leading cause of death, with $\ell^{\text{neoplasms}}(0) = 15,336$ on a radix of 100,000. If mortality from neoplasms were eliminated, e(0) would increase by 2.56 years. That increase is modest, in part because of the high death rates from cardiovascular diseases at ages 85 and over that limit life expectancy.

## 1.5 STABLE POPULATION MODELS

The stable population was pioneered by Alfred J. Lotka (1907, 1939), though the germ of the idea goes back to Euler (1760). While the stationary population reflects rates of decrement, the stable population reflects the long term implications of rates of both fertility and mortality. In doing so, it reveals the profound fact that constant rates of birth and death yield a population that grows exponentially over time, and that has an age composition that remains unchanged. The stable population was the dominant model of 20th century mathematical demography. By incorporating fertility and allowing growth, stable populations allow considerably more analytical flexibility than do stationary populations. Two important strengths of stable populations are that they provide a framework for analyzing populations and a basis for estimating demographic behavior in the absence of adequate data. Those estimates are often robust to substantial departures from the assumption of a history of constant vital rates. The discussion here draws on Caswell (2001), Keyfitz (1977), Pollard (1973), and Schoen (1988a).

### 1.5.1 Continuous Stable Models

Let us start with continuous functions and deal with female only populations. Demographers prefer female populations in fertility analyses because data on mothers are generally better than data on fathers, and the female reproductive age span is narrower and more clearly defined. It is clear that the number of births in a year result from the prevailing age-specific birth rates acting on the population of reproductive age, whose members are the survivors of births from earlier years. We denote by f(x,n) the constant model birth rate for daughters born to women between the ages of x to x + n, and let p(x) be the probability of surviving from birth to exact age x. Then if B(t) is the number of births in year t

$$B(t) = \int B(t-x)\, p(x)\, f(x)\, dx \qquad (1.27)$$

(Unless otherwise noted, all integrals range from 0 to infinity.) Sharpe and Lotka (1911) provided the basic solution for equation (1.27), though a rigorous and complete treatment had to await Feller (1941). Under restrictions always met by human patterns of fertility, the time trajectory of births has the form

$$B(t) = \sum_i Q_i \exp(r_i t) \qquad (1.28)$$

where the $Q_i$ are constants reflecting initial conditions (discussed in Chapter 2 and Appendix A) and the $r_i$ are growth rates, i.e. the instantaneous increase in population size per unit of population. Given the integral form of equation (1.27) there can be an unlimited number of roots, but it has been proven that there must be one unique real root that is larger than the other real roots and the real part of the complex roots. That dominant root, generally designated $r_1$ or simply r, has been termed the "intrinsic" growth rate or "Lotka's r". The other roots, being smaller than Lotka's r (and generally negative), become insignificant over time. Thus in the long term $B(t) = Q_1 \exp(r_1 t)$, and equation (1.27) can be rewritten and rescaled into

$$1 = \int \exp(-rx)\, p(x)\, f(x)\, dx \qquad (1.29)$$

Inspection of equation (1.29) shows that with mortality and fertility constant, a larger value of r must lower the value of the integral. As the integral takes on values from 0 to $\infty$, there can be one and only one real r that produces a value of 1.

Equation (1.29) is known as the characteristic or renewal equation, and expresses the core of stable population theory. It can be interpreted as showing that the unit birth in the reference year results from fertility at rate f(x) applied to the p(x) fraction of survivors to age x of the exp(−rx) births born x years earlier, summed over all ages x. The characteristic equation thus relates constant fertility and mortality to the stable, exponential birth sequence. It also provides an equation for Lotka's r in terms of that constant fertility and mortality.

The standard solution for r is due to Lotka (1939), and uses moments. Let us term p(x)f(x) the net maternity function, and express the $R_i$th moment of that function as

$$R_i = \int x^i\, p(x)\, f(x)\, dx, \quad i \geqslant 0 \qquad (1.30)$$

The zeroth moment, $R_0 = \int p(x) f(x)\, dx$, is the Net Reproduction Rate (or NRR), the number of daughters a woman would have in her lifetime under the given mortality and fertility rates. The ratio $R_1/R_0 = \mu$ gives the mean age at net maternity (or childbearing) in any stable population cohort. The variance of the net maternity function, $\sigma^2$, is given by $\{R_2/R_0 - [R_1/R_0]^2\}$. Expressing $e^{-rx}$ in equation (1.29) as a power series and stopping at second moments (thus assuming that net maternity can be adequately described by a normal curve), Lotka found the quadratic solution

$$r = \{\mu - [\mu^2 - 2\sigma^2 \ln R_0]^{1/2}\}/\sigma^2 \qquad (1.31)$$

A range of alternative solutions, yielding very similar values, is given in (Schoen, 1988a: Ch 3).

Given the logic underlying characteristic equation (1.29), we can find a number of stable population measures. The total number of persons in the stable population (in the reference year when there is one birth) is $T^*(0)$, that is

$$T^*(0) = \int \exp(-rx)\, p(x)\, dx \qquad (1.32)$$

or the sum of the survivors of all previous birth cohorts. The stable population crude (or intrinsic) birth rate, b, is then

$$b = 1/T^*(0) \tag{1.33}$$

The intrinsic death rate, or the crude death rate of the stable population, d, is

$$d = \int \exp(-rx)\, p(x)\, \mu(x)\, dx/T^*(0) \tag{1.34}$$

and by definition we have $r = b - d$. The fixed proportion of the population at age x last birthday, or the proportion attaining exact age x during any year, c(x), is

$$c(x) = \exp(-rx)\, p(x)/T^*(0) = b\, \exp(-rx)\, p(x) \tag{1.35}$$

From equation (1.32), it is easy to see that $\int c(x)\, dx = 1$. All of the above relationships apply in stationary populations, which are simply stable populations with $r = 0$.

Stable populations differ from stationary populations in one important respect: period measures in a stable population are not the same as cohort measures. The behavior of every stable population cohort is the same; only its size is different. Every period population is also the same, apart from size. Each period population, however, reflects a combination of cohorts of different size. This if $r > 0$, then the mean age of the stable population, $Y_r$, is less than the mean age of the stationary population based on the same survivorship, because younger cohorts are larger than older cohorts. Mathematically,

$$Y_r = \int x\, \exp(-rx)\, p(x)\, dx/T^*(0) \tag{1.36}$$

The larger the value of $r > 0$, the more younger cohorts are weighted relative to older cohorts. With respect to mortality, d is less than $1/e(0)$ when $r > 0$, as the life table crude death rate reflects the older stationary population.

The stable population mean age at childbearing, $A^*$, differs from the cohort measure $\mu$ and can be written

$$A^* = \int x\, \exp(-rx)\, p(x)\, f(x)\, dx \tag{1.37}$$

as the denominator is the characteristic equation and thus equals 1. Lotka introduced a third measure of the timing of fertility in stable populations, the mean length of a generation, $\mp$. It reflects the length of time needed for a stable population to grow by a factor equal to its NRR, or

$$R_0 = \exp(r\mp) \tag{1.38}$$

The three measures, $\mu$, $A^*$, and $\mp$ are identical when $r = 0$. When $r \neq 0$, $\mp$ is close to the geometric mean of $\mu$ and $A^*$.

### 1.5.2 Discrete Stable Models

Discrete functions are always involved when continuous population relationships are applied to data. Beyond that, the discrete formulation affords a distinct and valuable perspective on stable population behavior.

In terms of n year age intervals, characteristic equation (1.29) can be written

$$1 = \sum \exp(-r[x+n/2])\, L(x,n)\, f(x,n)/\ell(0) \tag{1.39}$$

where the summation over the life span proceeds in steps of n years. Keyfitz (1977) used that relationship to devise an iterative solution for Lotka's r. Multiplying both sides of equation (1.39) by exp[27.5r], which introduces r on the left hand side and improves the rate of convergence, taking the natural logarithm, and then rewriting gives

$$r' = (1/27.5)\ \ln\ \{\Sigma \exp(-r[x-27.5+n/2])\, L(x,\,n)\, f(x,\,n)/\ell(0)\} \tag{1.40}$$

where improved intrinsic growth rate r′ is obtained from previously calculated value r. Starting the iterative process with $r = 0$, convergence to 6 decimals is generally achieved in about 6 iterations.

Discrete representations of stable populations stem from the works of Bernardelli (1941), Lewis (1942), and Leslie (1945). As Leslie (1945) gave the most thorough treatment, and published in the most visible journal, the matrix model is generally associated with his name. We assume interval lengths of n years, which divide the ages below the maximum age at reproduction (age 45 or 50) into $\gamma$ intervals. Consider the $\gamma \times \gamma$ matrix

$$\mathbf{A} = \begin{bmatrix} b_1 & b_2 & b_3 & \dots & b_{\gamma-1} & b_\gamma \\ s_1 & 0 & 0 & \dots & 0 & 0 \\ 0 & s_2 & 0 & \dots & 0 & 0 \\ & & \cdot & & & \\ & & & \cdot & & \\ & & & & \cdot & \\ 0 & 0 & 0 & \dots & s_{\gamma-1} & 0 \end{bmatrix} \tag{1.41}$$

where the $b_i$ in the first row reflect fertility, the $s_i$ on the subdiagonal reflect survivorship, and all of the other matrix elements are zero. More precisely, $b_i$ reflects the contribution of persons in the ith age group at initial time t to the number of persons in the first age group at the end of the interval, time $t+n$. Survivorship ratios $s_i$ reflect the proportion of persons in the ith age group at initial time t who survive to be in the $(i+1)$st age group at time $t+n$ [see Exercise 13, Section 1.9].

Leslie matrix $\mathbf{A}$ is a population projection matrix (PPM), that is it can project an initial population of reproductive age n years ahead. Let $\mathbf{x_t}$ be a $\gamma \times 1$ column vector whose ith element, $x_{it}$, is the number of persons in the ith age group at time t. We then have the relationship

$$\mathbf{x_{t+n}} = \mathbf{A}\,\mathbf{x_t} \tag{1.42}$$

The sum of the $b_i x_{it}$ yields the number of persons in the first age group at time $t+n$, that is the number of births during the t to $t+n$ interval who survive to time $t+n$. The subdiagonal elements provide the number of persons in age groups 2 through $\gamma$. The probability of surviving from age group 1 to age group i, $p_i$, can then be written

$$p_{i+1} = \prod_{i=1}^{i} s_i \tag{1.43}$$

with $p_1 = 1$.

A stable population can be characterized by a Leslie matrix that is constant over time, and the structure of that Leslie matrix affords another way to find the stable growth rate and age composition. Appendix A gives a brief discussion of the eigenstructure of a matrix, i.e. the unique decomposition that can be performed on a broad range of matrices. That decomposition is especially important in demographic work because its results have substantive interpretations. The dominant eigenvalue of $\mathbf{A}$, or its dominant root, is generally denoted by $\lambda_1$ or just $\lambda$. It is the stable growth parameter (often called the stable growth rate, though that departs from our usual use of the term rate) over the n year interval, hence

$$\lambda = \exp(rn) \tag{1.44}$$

The dominant right eigenvector of $\mathbf{A}$ is the $\gamma \times 1$ column vector $\mathbf{u}$, whose first element is 1 and whose ith element is $u_i$. In the stable population, $u_i$ describes the age composition, as it gives the size of the ith age group relative to the first age group. Algebraically, for $i > 1$, $u_i = p_i \lambda^{1-i}$, a discrete representation analogous to $c(x)/c(0)$ from equation (1.35). The Leslie matrix underlying a stable population thus provides the basic parameters of that stable population. If we have a stable population with one person in the first age group at time t, equation (1.42) can be rewritten

$$\lambda\,\mathbf{u} = \mathbf{A}\,\mathbf{u} \tag{1.45}$$

Projecting the initial stable population over one interval has the effect of increasing the number at every age by a factor of $\lambda$.

## 1.6 MULTISTATE LIFE TABLES

Multistate or increment-decrement life tables extend the multiple decrement model by allowing cohort survivors to be followed through different living states. A multistate life table describes the life course of a cohort as its members age and move through the specified set of states. The state space of the model can be chosen to suit the needs of the analyst, as the model states need only be exhaustive and mutually exclusive. The first multistate table is apparently due to L.G. DuPasquier (1912), who analyzed disability insurance using a model with the states Healthy and Disabled. Despite later work by Fix and Neyman (1951) and Chiang (1964), multistate models received little attention until the 1970s, but by the early 1980s their relevance to demography was well established (cf. Land and Rogers 1982). Multistate life tables have proven to be extremely useful in demographic work (cf. Schoen 1988b), and have been applied to analyses of marital and union statuses, labor force behavior, interregional migration, and numerous other topics. The discussion that follows draws substantially on Schoen (1988a).

### 1.6.1 Scalar Multistate Equations

Consider a multistate model with k living states, at least two of which are connected by decrements. Using the General Algorithm, the model can be specified, age by age, by a series of flow, orientation and person-year equations. Let $\ell_i(x)$ be the model number of persons in state i at exact age x, and let $d_{ij}(x, n)$ be the number of decrements (or moves or transfers) from state i to state j between the ages of x and x + n. The model has k flow equations of the form

$$\ell_i(x+n) = \ell_i(x) - \Sigma_{j\neq i}\, d_{ij}(x, n) + \Sigma_{j\neq i}\, d_{ji}(x, n) \qquad (1.46)$$

where the flows include decrements to absorbing (or "dead") state k + 1. In short, the number of persons in state i at exact age x + n equals the number in state i at exact age x, minus the decrements to, and plus the increments from, all of the other states in the model. The ratio $\ell_i(x+n)/\ell_i(x)$ can exceed 1 because of increments from other states, and hence is not a "proper" probability. The population based or "p-probability", $p_i(x, n)$, is

$$p_i(x, n) = \ell_i(x+n)/\ell(x) \qquad (1.47)$$

and represents the probability that a person alive at exact age x is in state i at exact age x + n.

Equation (1.46) deals with "net flows", that is $\ell_i(x)$ functions that only specify the state occupied at age x. However, it is also important to consider "gross flow" functions of the form $\ell_{ij}(x+u)$, which represent the number of persons in state j at exact age x+u who were also in state i at exact age x. Similarly, let $d_{ijh}(x, n)$ be the number of moves from state j to state h between the ages of x and x+n by persons in state i at exact age x. At the level of the gross flows, there are $k^2$ flow equations of the form

$$\ell_{ij}(x+n) = \ell_{ij}(x) - \Sigma_{h \neq j}\, d_{ijh}(x, n) + \Sigma_{h \neq j}\, d_{ihj}(x, n) \qquad (1.48)$$

where $\ell_{ij}(x)$, $i \neq j$, is zero, as no one can be in two states at one time. Note that from the way $\ell_{ij}(x+u)$ is defined, the "+" is part of the function definition, and indicates more than addition. If $z = x+u$, we do *not* have $\ell_{ij}(x+u) = \ell_{ij}(z)$. After the $\ell_{ij}(x+n)$ have been determined, they can be summed over i to provide the initial $\ell_j$ value for the next age interval.

The gross flow functions give rise to the Markov transition probabilities, or "$\pi$-probabilities". Specifically, $\pi_{ij}(x, n)$, the probability that a person in state i at exact age x is in state j at exact age x+n is given by

$$\pi_{ij}(x,n) = \ell_{ij}(x+n)/\ell_i(x) \qquad (1.49)$$

Following our usual procedure, we assume that the model reproduces the observed rates. Let $M_{ij}(x, n)$ be the observed rate of decrement from state i to state j between the ages of x and x+n, and let $m_{ij}(x, n)$ be the analogous model rate. We then have $k^2$ orientation equations of the form

$$M_{ij}(x, n) = m_{ij}(x, n) \qquad (1.50)$$

where $m_{ij}(x, n) = d_{ij}(x, n)/L_i(x, n)$, and $L_i(x, n)$ represents the number of person-years lived in state i between the ages of x and x+n. For the rates, there is no need to recognize the state occupied at the beginning of the interval. The model employs the traditional Markov assumption that past history does not matter, and that the risk of movement depends only on present state and age. The underlying force of transition, or instantaneous rate of transfer, from state i to state j at exact age x, $\mu_{ij}(x)$, is defined as

$$\mu_{ij}(x) = \lim_{u \downarrow 0} \pi_{ij}(x, u)/u \qquad , i \neq j \qquad (1.51)$$

The person-year equations finish the specification of the model. At the level of the net flows, we have k equations of the form

$$L_i(x, n) = \int_0^n \ell_i(x+u)\, du \qquad (1.52)$$

and at the level of the gross flows, $k^2$ equations of the form

$$L_{ij}(x, n) = \int_0^n \ell_{ij}(x+u)\, du \qquad (1.53)$$

where $L_{ij}(x, n)$ represents the number of person-years lived in state j between the ages of x and x + n by persons who were in state i at exact age x.

### 1.6.2 Matrix Multistate Equations

As with the basic life table, model construction starts at age 0 and proceeds age by age. Finding successive values of $\ell_{ij}(x+n)$ is greatly facilitated by the introduction of matrices. Let the end of the age interval gross flow matrix, $\boldsymbol{\ell}(\mathbf{x}+\mathbf{n})$, be

$$\boldsymbol{\ell}(\mathbf{x}+\mathbf{n}) = \begin{bmatrix} \ell_{11}(x+n) & \ell_{12}(x+n) & \dots & \ell_{1,k+1}(x+n) \\ \ell_{21}(x+n) & \ell_{22}(x+n) & \dots & \ell_{2,k+1}(x+n) \\ \vdots & \vdots & & \vdots \\ \ell_{k+1,1}(x+n) & \ell_{k+1,2}(x+n) & \dots & \ell_{k+1,k+1}(x+n) \end{bmatrix} \qquad (1.54)$$

where we include a (k + 1)st "dead" state that persons enter but do not leave. Diagonal matrix $\boldsymbol{\ell}(\mathbf{x})$ is also specified by equation (1.54), as $\ell_{11}(x) = \ell_1(x)$.

The matrix of decrements, **d(x,n)**, can be written

$$\mathbf{d}(\mathbf{x}, \mathbf{n}) = \begin{bmatrix} \sum[d_{11j} - d_{1j1}] & \sum[d_{12j} - d_{1j2}] & \dots & \sum[-d_{1j,k+1}] \\ \sum[d_{21j} - d_{2j1}] & \sum[d_{22j} - d_{2j2}] & \dots & \sum[-d_{2j,k+1}] \\ \vdots & \vdots & & \vdots \\ 0 & 0 & \dots & 0 \end{bmatrix} \qquad (1.55)$$

where the (x,n) interval identifiers have been suppressed. From equations (1.54) and (1.55) we have the matrix flow equation

$$\boldsymbol{\ell}(\mathbf{x}+\mathbf{n}) = \boldsymbol{\ell}(\mathbf{x}) - \mathbf{d}(\mathbf{x}, \mathbf{n}) \qquad (1.56)$$

which is analogous to basic life table equation (1.7).

The matrix of transition probabilities, $\mathbf{\Pi}(\mathbf{x}, \mathbf{n})$, can be written

$$\mathbf{\Pi}(\mathbf{x}, \mathbf{n}) = \begin{bmatrix} \pi_{11}(x, n) & \pi_{12}(x, n) & \dots & \pi_{1,k+1}(x, n) \\ \pi_{21}(x, n) & \pi_{22}(x, n) & \dots & \pi_{2,k+1}(x, n) \\ \cdot & \cdot & & \cdot \\ \cdot & \cdot & & \cdot \\ \cdot & \cdot & & \cdot \\ \pi_{k+1,1}(x, n) & \pi_{k+1,2}(x, n) & \dots & \pi_{k+1,k+1}(x, n) \end{bmatrix} \tag{1.57}$$

From equations (1.54) and (1.57), it can be seen that

$$\Pi(\mathbf{x}, \mathbf{n}) = \boldsymbol{\ell}^{-1}(\mathbf{x})\boldsymbol{\ell}(\mathbf{x}+\mathbf{n}) \tag{1.58}$$

where the superscript $(-1)$ indicates the matrix inverse.

The matrix of observed or data transition rates, $\mathbf{M(x,n)}$, can be written

$$\mathbf{M}(\mathbf{x}, \mathbf{n}) = \begin{bmatrix} \sum M_{1j}(x, n) & -M_{12}(x, n) & \dots & -M_{1,k+1}(x, n) \\ -M_{21}(x, n) & \sum M_{2j}(x, n) & \dots & -M_{2,k+1}(x, n) \\ \cdot & \cdot & & \cdot \\ \cdot & \cdot & & \cdot \\ \cdot & \cdot & & \cdot \\ 0 & 0 & \dots & 0 \end{bmatrix} \tag{1.59}$$

Note that each row sums to zero. The matrix of model rates, $\mathbf{m(x,n)}$, has the same structure, but contains m instead of M rates. Our orientation equations are then expressed by

$$\mathbf{M(x,n) = m(x,n)} \tag{1.60}$$

The person-year matrix, $\mathbf{L(x,n)}$, has the same structure as $\boldsymbol{\ell}(\mathbf{x}+\mathbf{n})$, and its (i,j)th element is $L_{ij}(x, n)$. Paralleling basic life table equation (1.9)

$$\mathbf{L}(\mathbf{x}, \mathbf{n}) = \int_0^n \boldsymbol{\ell}(\mathbf{x}+\mathbf{u})\, \mathbf{du} \tag{1.61}$$

It follows from the definitions of the $\mathbf{L}$, $\mathbf{d}$, and $\mathbf{m}$ matrices that

$$\mathbf{d}(\mathbf{x}, \mathbf{n}) = \mathbf{L}(\mathbf{x}, \mathbf{n})\, \mathbf{m}(\mathbf{x}, \mathbf{n}) \tag{1.62}$$

the matrix analog to basic life table equation (1.10). The flow, orientation, and person-year equations of the multistate model can thus be written compactly in matrix form in a way that closely resembles their basic life table counterparts.

The three life table solutions discussed in section 1.3 can be extended to the multistate case, again using matrix equations that parallel the previous scalar expressions. The exponential or constant forces solution is

$$\boldsymbol{\ell}(\mathbf{x}+\mathbf{n}) = \boldsymbol{\ell}(\mathbf{x})\,\mathbf{exp}[-\mathrm{n}\mathbf{M}(\mathbf{x},\mathbf{n})] \tag{1.63}$$

That exponentiated matrix is defined by an infinite power series, i.e.

$$\exp[-\mathrm{n}\mathbf{M}(\mathbf{x},\mathbf{n})] = \mathbf{I} - \mathrm{n}\mathbf{M}(\mathbf{x},\mathbf{n}) + (\mathrm{n}^2/2!)\mathbf{M}^2(\mathbf{x},\mathbf{n}) - (\mathrm{n}^3/3!)\mathbf{M}^3(\mathbf{x},\mathbf{n}) + \ldots \tag{1.64}$$

where $\mathbf{I}$ is the $(k+1) \times (k+1)$ identity matrix. The series always converges, though not necessarily rapidly.

The linear multistate solution can be written

$$\boldsymbol{\ell}(\mathbf{x}+\mathbf{n}) = \boldsymbol{\ell}(\mathbf{x})[\mathbf{I} - (\mathrm{n}/2)\mathbf{M}(\mathbf{x},\mathbf{n})][\mathbf{I} + (\mathrm{n}/2)\mathbf{M}(\mathbf{x},\mathbf{n})]^{-1} \tag{1.65}$$

The more general Mean Duration at Transfer approach has gross flows given by

$$\boldsymbol{\ell}(\mathbf{x}+\mathbf{n}) = \boldsymbol{\ell}(\mathbf{x})[\mathbf{I} - \mathbf{U}(\mathbf{x},\mathbf{n})\mathbf{M}(\mathbf{x},\mathbf{n})][\mathbf{I} + \mathrm{n}\mathbf{M}(\mathbf{x}+\mathbf{n}) + \mathbf{W}(\mathbf{x},\mathbf{n})\mathbf{M}(\mathbf{x},\mathbf{n})]^{-1} \tag{1.66}$$

where the **U** and **W** are weight matrices with elements paralleling the u(x,n) and w(x,n) functions defined after equation (1.18). Specifically, the (i,j)th element of **U** and **W** is that weighted function of $M_{ij}$ values.

The linear matrix of person-year values is given by

$$\mathbf{L}(\mathbf{x},\mathbf{n}) = (\mathrm{n}/2)[\boldsymbol{\ell}(\mathbf{x}) + \boldsymbol{\ell}(\mathbf{x}+\mathbf{n})] \tag{1.67a}$$

and the MDAT person-year matrix by

$$\mathbf{L}(\mathbf{x},\mathbf{n}) = \mathrm{n}\boldsymbol{\ell}(\mathbf{x}+\mathbf{n}) + \boldsymbol{\ell}(\mathbf{x})\mathbf{U}(\mathbf{x},\mathbf{n}) + \boldsymbol{\ell}(\mathbf{x}+\mathbf{n})\mathbf{W}(\mathbf{x},\mathbf{n}) \tag{1.67b}$$

The elements of the exponential person-year matrix parallel equation (1.13).

Special procedures are again needed at the youngest and oldest ages. Infant deaths occur, on average, well before age ½, but the linear approximation may be satisfactory for other causes of decrement (e.g. interregional migration). To end the table, one can use the matrix analog of equation (1.20), that is

$$\mathbf{L}(\mathbf{90},\infty) = \boldsymbol{\ell}(\mathbf{90})\mathbf{M}^{-1}(\mathbf{90},\infty) \tag{1.68}$$

Since the $(k+1)$ by $(k+1)$ **M** matrix of equation (1.59) has a row of zeroes and cannot be inverted, a $k \times k$ **M** matrix that excludes the $(k+1)$st row and column should be used in equation (1.68).

The equations presented above assume that data are available in the form of occurrence/exposure rates. That is often not the case, but multistate models can be constructed from other forms of data as well. Event history data generally allow the determination of events and the person-years of exposure lived in each state. Prospective and even retrospective data can be used, as they allow the calculation of survivorship proportions.

Let the data survivorship proportion $S_{ij}(x, n)$ represent the proportion of persons in state i between the ages of x and x + n who are in state j exactly n years later. (If the data are retrospective, the size of the earlier population has to be adjusted for loss to mortality over the interval; that is generally done by using a relevant life table and assuming the same mortality in all states.) Collect the S proportions in matrix $\mathbf{S(x,n)}$, whose (i,j)th element is $S_{ij}(x, n)$. The **S** matrices can then be related to the multistate transition probability matrices in several ways. Rees and Wilson (1977) used the relationship

$$\boldsymbol{\Pi}(\mathbf{x}, \mathbf{n}) = {}^{1}/_{2}[\mathbf{S}(\mathbf{x}, \mathbf{n}) + \mathbf{S}(\mathbf{x} - \mathbf{n}, \mathbf{n})] \qquad (1.69)$$

and Hoem and Fong (1976) proposed

$$\boldsymbol{\Pi}(\mathbf{x} + \mathbf{n}/2, \mathbf{n}) = \mathbf{S}(\mathbf{x}, \mathbf{n}) \qquad (1.70)$$

Both of those assumptions are essentially linear, so the linear solution of equation (1.65) and general equations (1.56), (1.58), and (1.61) can be used to find the remaining functions. [Exercise 10 in Section 1.9 provides relationships between $\boldsymbol{\Pi}(\mathbf{x} + \mathbf{n}/2, \mathbf{n})$ and $\boldsymbol{\Pi}(\mathbf{x}, \mathbf{n})$.]

### 1.6.3 Multistate Expectancies and Summary Measures

Life expectancy measures are essentially integrals, or sums, of probabilities. Consequently, the p- and $\pi$-probabilities lead to different measures of average future lifetimes.

The p-probabilities lead to *population based* expectations of life. Let $e_i(x)$ denote the average future lifetime a person exact age x can expect to spend in state i. Then

$$e_i(x) = \int \ell_i(x + u)\, du / \ell(x) \qquad (1.71)$$

That integral is the sum of the $L_i$ values from age x to the end of the table, and hence is readily calculable. Population based life expectancies are useful in many analyses, providing such measures as the expected number of years a person age 15 will spend in the labor force or the expected number of years a person age 50 will spend disabled.

The $\pi$-probabilities lead to *status based* life expectancies that are conditional on status at initial age x. Let $\varepsilon_{ij}(x)$ be the status based expectancy in state j for a person exact age x who is in state i. Then

$$\varepsilon_{ij}(x) = \int \ell_{ij}(x+u)\,du/\ell_i(x) \qquad (1.72)$$

Because $\varepsilon_{ij}(x)$ is based on the closed group of persons in state i at age x, it cannot be found from conventional multistate life table functions, where the closed group is respecified at the beginning of each interval. A supplementary calculation must be made to determine the future life course of a person age x in state i. Status based life expectancies are valuable measures and often differ markedly from population based expectancies. For example, the average number of years a person age 50 will spend in the Married state is generally quite different for persons who are married and persons who are not.

Life expectancies are only one of many summary measures of behavior that can be found in multistate models. Analyses may be interested in the proportion of life lived by state, in the average age of the population by state, in the number of decrements between specified states, and in the average age at each type of transfer. Different substantive areas tend to emphasize different measures. In labor force status models, the average number of labor force entries (or exits) per person is often of interest. In a multiregional model where all persons are born in a particular region, specifics regarding movement patterns to other states are important. In marital status life tables, the model can yield the probability a marriage ends in divorce and the average length of a marriage, measures difficult to obtain in other ways.

Multistate models with any state space can be calculated from the matrix equations given above by means of mathematical packages such as STATA, S-PLUS, MATLAB, MAPLE, or MATHEMATICA. A number of specific computer programs are also available to simplify the task. Appendix D to Schoen (1988a) provides FORTRAN programs for a two living state model based on survivorship proportions and for a four living state model based on rates. Tiemeyer and Ulmer (1991) have a program in C++ that can accommodate any number of states, and Rogers (1995) provides a DOS compatible program.

## 1.7 THE MULTISTATE STABLE POPULATION

The multistate stable population was pioneered by Rogers (1975), and can combine the three core processes of demography. It shows that the long term implications of rates of mortality, fertility, and interstate transfer are a population with an unchanging age-state composition that is growing exponentially at a constant, intrinsic, rate r.

The characteristic or renewal equation for a multistate stable population can be given by

$$\boldsymbol{\ell}(\mathbf{0}) = \int \exp(-rx)\, \boldsymbol{\ell}(\mathbf{x})\, \mathbf{f}(\mathbf{x})\, \mathbf{dx} \qquad (1.73)$$

where the $(k+1) \times (k+1)$ diagonal matrix $\mathbf{f(x)}$ has $f_i(x)$, the fertility rate in state i at age x, as its ith diagonal element. In the multistate life table (or multistate stationary population), the age zero state composition can be chosen by the analyst. In the multistate stable population, however, it is determined by the constant rates of fertility, mortality, and interstate transfer. Given a unit total birth cohort size in the reference year, matrix equation (1.73) represents k equations for k unknowns: intrinsic r and the relative sizes of the number of births in the k living states. For state j, then, equation (1.73) implies the scalar equation

$$\ell_j(0) = \int \exp(-rx)\, \ell_j(x)\, f_j(x)\, dx \qquad (1.74)$$

where $f_j(x)$ is the fertility rate in state j (for births in state j) at age x. The number of births in state j thus results from the application of fertility rate $f_j(x)$ to the $\ell_j(x)$ fraction of births, from a cohort of size $e^{-rx}$, who survived to be in state j at age x, summed over all ages x. The unit birth in the reference year is the sum of the births in all k living states.

Finding intrinsic growth rate r and the age zero state composition is best accomplished by functional iteration (cf. Schoen 1988a). Extending the approach underlying equations (1.39) and (1.40), we can write k discrete equations of the form

$$\ell_i(0) = \sum \exp(-r[x+n/2])\, L_i(x,n)\, f_i(x,n) \qquad (1.75)$$

where the summation over age proceeds in steps of n years from 0 to the highest age of reproduction. Summing those k equations over all states, multiplying by $e^{27.5r}$, taking logs and rewriting leads to an iterative equation for the intrinsic growth rate, i.e.

$$r' = (1/27.5) \ln \{\Sigma_i \Sigma_x \exp(-r[x-27.5+n/2])\, L_i(x,n)\, f_i(x,n)/\ell(0)\} \qquad (1.76)$$

Start with $r = 0$ and a multistate life table to find the first improved r' value. That improved r' is used in equation (1.75) to provide better values for all of the $\ell_i(0)$. Those updated $\ell_i(0)$ values are then used to calculate a new multistate life table, which yields improved $L_i$ values. In turn, those $L_i$ values are employed in equation (1.76) to yield a further improved r′. The number of iterations required depends on the specified tolerance between r and r′ and on how close the initial $\ell_i(0)$ are to the final multistate distribution.

The General Algorithm for constructing a multistate stable population adds renewal equation (1.73) to the flow, orientation, and person-year equations of the previous section. The orientation equations now include the specification that the data fertility rate in state i at age x, $F_i(x)$, equals its model counterpart $f_i(x)$.

The total number of persons in the multistate stable population in the year when there is one birth, $T^{**}(0)$, is

$$T^{**}(0) = \Sigma_i \int \exp(-rx)\, \ell_i(x)\, dx \qquad (1.77)$$

where the sum ranges over the k living states. Equation (1.77) is the multistate version of equation (1.32). The intrinsic birth rate, b, is

$$b = 1/T^{**}(0) \qquad (1.78)$$

If we denote by $c_i(x)$ the proportion of the multistate stable population in state i at age x, or the proportion that attains age x in state i, then

$$c_i(x) = b \exp(-rx)\, \ell_i(x) \qquad (1.79)$$

with $c(x) = \Sigma_i\, c_i(x)$.

The multistate stable population has received only modest use. The model has extensive data requirements and is not simple to construct. The assumption of constant interstate transfer rates is often difficult to justify. In addition, the age-state composition of the multistate stable population appears to have limited analytical value. As we see in Chapter 2, multistate stability can take considerably longer to arise than does stability in birth-death models.

## 1.8 SUMMARY

This chapter provides a brief introduction to fixed rate population models. Procedures for life table construction are set out in a General Algorithm, whose flow, orientation, and person-year equations specify the model and the relationship between survivorship ($\ell$) and person-year (L) values. For stable population model construction, an equation for population renewal is also required.

The basic life table describes the experience of a cohort from birth to the death of its last member. The life table can also be interpreted as a stationary population, and as such provides a model with a constant size and age composition. Recognizing causes of decrement leads to multiple decrement models, and to their associated, or cause eliminated, variants.

The stable population model is based on the fact that constant rates of birth and death lead, in time, to a population that grows exponentially and has a fixed age composition. The stable model provides a framework for population analysis and a tool to estimate demographic measures in the context of limited data.

Multistate life tables recognize any number of living states, and allow cohort members to move between them in prescribed ways. They provide a detailed description of cohort behavior with respect to the states in the model. Multistate stable populations add growth, and provide a model that can capture the long term implications of fertility, mortality, and migration, the three core process of demography.

## 1.9 EXERCISES

[See Appendix B for selected answers]

1. Show that in a cohort's last year of life
   a. the death rate (M) exceeds 1
   b. the force of mortality ($\mu$) exceeds 1
2. The Expectation of Productive Life (EPL) has been defined as the number of years a person age 15 will live between the ages of 15 and 65. Express the EPL in terms of life table functions.
3. In terms of the death rate in the open ended interval $M(90, \infty)$, what is
   a. e(90)
   b. $a(90, \infty)$
4. In 1724, de Moivre suggested representing the survivorship curve by a straight line. Jordan (1975) expressed such a curve as $\ell(x) = k\ (\omega - x)$, where $\omega$ is the highest age attained by the cohort. Under that form of survivorship, what is
   a. $\mu(x)$
   b. e(x)
5. The most famous Law of Mortality is that proposed by Benjamin Gompertz in 1825, which expresses the force of mortality as an exponential function. Under Gompertz' Law, we can write $\mu(x) = \exp(A + bx)$. Show that implies $\ell(x) = \ell(0) \exp[(e^A - e^{A+bx})/b]$.
6. Consider a double decrement table with causes of death i and (−i) [read "not i"]. In the ASDT with cause i active, the probability of surviving to age x is $p^{(i)}(x)$, and in the ASDT with cause (−i) active, the probability of surviving to age x is $p^{(-i)}(x)$. If the probability of surviving to age x in the double decrement life table is p(x), show that $p(x) = p^{(i)}(x)p^{(-i)}(x)$.
7. In a certain stationary population, everyone between the ages of 20 and 65 works, while everyone 65 and over is retired. If all retired persons receive a pension of $D a year, how much must each person 15 to 65 contribute to provide that benefit?

8. Show that in a stable population with B births in the reference year ($t = 0$), the number of persons attaining age x in year t, N(x,t), equals $B\exp[-r(x-t)]p(x)$.
9. In a stable population, assume that d and the p(x) schedule are known. Find r, b and c(x).
10. In a linear multistate life table, show that
    a. $\mathbf{M}(\mathbf{x}, \mathbf{n}) = (2/n)\,[\mathbf{I}+\boldsymbol{\Pi}(\mathbf{x}, \mathbf{n})]^{-1}\,[\mathbf{I}-\boldsymbol{\Pi}(\mathbf{x}, \mathbf{n})]$
    b. $\mathbf{S}(\mathbf{x}, \mathbf{n}) = [\mathbf{I}-(n/2)\mathbf{M}(\mathbf{x}, \mathbf{n})]\,[\mathbf{I}+(n/2)\mathbf{M}(\mathbf{x}+\mathbf{n}, \mathbf{n})]^{-1}$
    c. $\boldsymbol{\Pi}(\mathbf{x}, \mathbf{n}/2) = [\mathbf{I}+\boldsymbol{\Pi}(\mathbf{x}, \mathbf{n})]/2$
    d. $\boldsymbol{\Pi}(\mathbf{x}+\mathbf{n}/2, \mathbf{n}/2) = 2\,[\mathbf{I}+\boldsymbol{\Pi}(\mathbf{x}, \mathbf{n})]^{-1}\,\boldsymbol{\Pi}(\mathbf{x}, \mathbf{n})$
11. In a two living state linear multistate life table, show that

$$\pi_{12}(x, n)/\pi_{21}(x, n) = M_{12}(x, n)/M_{21}(x, n)$$

12. Miller (1946) presented a $2n+1$ term weighted moving average to produce graduated $u_x$ values from observed $u''_x$ values using the relationship

$$u_x = \Sigma\, a_{|n-i|}\, u''_{x+n-i}$$

where the sum over i ranges from 0 to 2n and the vertical bars indicate the absolute value. The a's are weights that sum to 1, with $a_r = a_{-r}$, $r = 1..n$. Taking a cubic polynomial as the standard for smoothness, the maximum reduction in mean square error is obtained when the $a_r$ are given by

$$a_r = \{3(3\,n^2+3n-1)-15\,r^2\}/\{(2n-1)(2n+1)(2n+3)\} \qquad , r = 0..n$$

With $u''_x$ values of 13, 15, 20, 18 and 19, use a 5 term moving average to "smooth" the middle $u''_x$ value.
13. Show that $b_j$, the jth element of the first row of a Leslie matrix, can be written as

$$b_j = Z(f_j + s_j\, f_{j+1})/2$$

where $f_j$ is the fertility of the jth age group, $s_j$ survives persons from the jth to the $(j+1)$st age group, and Z is the number of life table person years lived in the (0,n) age interval divided by $\ell(0)$. If $f_j$ is not restricted to female births, Z needs to include a factor reflecting the fraction of births that are female. [Note: If $L_j$ represents the number of person years lived in the jth age group, $s_j = L_{j+1}/L_j$.]
14. Prove that the age composition of any observed population can be represented as the age composition of a stable composition (though not necessarily a realistic one). [*Hint*: use the **u** (relative age composition) eigenvector.]

15. Regarding Chiang's a, show that equations (1.17) and (1.18) imply that

$$a(x, n) = \frac{u(x, n) + w(x, n) + nM(x, n)u(x, n)}{M(x, n)[n + u(x, n) + w(x, n)]}$$

where u(x,n) and w(x,n) are as defined after equation (1.18).

# CHAPTER 2

# CONVERGENCE TO STABILITY

## 2.1 INTRODUCTION

The experience of a cohort spans a single lifetime, but the emergence of a stable (or stationary) population can take considerably longer. In this chapter, we examine the conditions under which a time invariant set of rates produces a population with a fixed age (or state) composition, and explore the process of convergence to stability.

## 2.2 CONDITIONS FOR STABILITY

Human populations subject to fixed rates generally find the population composition that reproduces itself under those rates, though the mathematics of convergence is a bit more complicated. The discussion in this section emphasizes human population models, drawing on Caswell (2001, Ch 4) and the matrix algebra presented in Appendix A.

A $\gamma \times \gamma$ population projection matrix, whether it be a Leslie matrix for a birth-death model or a population projection matrix (PPM) for a multistate population, has $\gamma$ eigenvalues. The nature of those eigenvalues determines whether convergence to a fixed stable model occurs and whether the initial composition of the population is forgotten. Let our PPM, $\mathbf{A}$, be written in eigenstructure form [equation (A.3)] as

$$\mathbf{A} = \mathbf{U}\boldsymbol{\Lambda}\mathbf{V} \tag{2.1}$$

where $\mathbf{U}$ is the matrix of right eigenvectors, $\mathbf{V} = \mathbf{U}^{-1}$, and $\boldsymbol{\Lambda}$ is the diagonal matrix of eigenvalues, with dominant eigenvalue $\lambda_1$ the (1,1) element. Projecting t intervals ahead is equivalent to raising $\mathbf{A}$, or $\boldsymbol{\Lambda}$, to the t-th power, i.e.

$$\mathbf{A}^t = \mathbf{U}\boldsymbol{\Lambda}^t\mathbf{V} \tag{2.2}$$

The process of convergence is essentially the process where $\boldsymbol{\Lambda}^t$ becomes a scalar, $\lambda_1^t$, as the subordinate eigenvectors become zero (or insignificant relative to $\lambda_1^t$). Under certain circumstances, however, that does not happen.

As we are dealing with nonnegative PPMs, i.e. demographic matrices that cannot have negative elements, the Theorem of Perron and Frobenius tells us that there are three possible cases: **A** is primitive, imprimitive, or reducible. In all cases, the dominant root, $\lambda_1$, is nonnegative. The most common case demographically is for **A** to be primitive. A Leslie matrix is primitive if any two adjacent first row elements are nonzero, or if any two nonzero first row elements are in positions (1,i) and (1,j) where i and j are relatively prime. Thus a $4 \times 4$ Leslie matrix is primitive if there are positive numbers in first row positions 1 and 2 or 2 and 3, but not if the only nonzero elements are in positions 2 and 4. In general, a $\gamma \times \gamma$ PPM **A** is primitive if all of its elements are positive (i.e. nonzero), or if $\mathbf{A}^{\gamma(\gamma-2)+2}$ has all positive elements.

A primitive matrix converges to stability, as there is some time T such that the subordinate eigenvectors can be ignored and **A** becomes a rank one matrix, that is a matrix that can be written as the product of two vectors. At time T, equation (2.2) reduces to

$$\mathbf{A}^{\mathrm{T}} = \lambda_1{}^{\mathrm{T}}\mathbf{u}\mathbf{v}' \tag{2.3}$$

where **u** is the dominant right eigenvector (the first column of **U**, which is associated with $\lambda_1$) and **v**' is the dominant left eigenvector (the first row of **V**).

Let the population vector at time T be $\mathbf{x}_\mathrm{T}$, whose ith element $x_{i\mathrm{T}}$ is the number of persons in group i at time T. Projecting the initial population to time T (see equation (1.42)) yields

$$\mathbf{x}_\mathrm{T} = \lambda_1{}^{\mathrm{T}}\mathbf{u}(\mathbf{v}'\mathbf{x}_0) \tag{2.4}$$

where $\lambda_1$ is the stable growth parameter and **u** represents the stable state composition. The product $(\mathbf{v}'\mathbf{x}_0)$ is a scalar, often designated by Q, and represents the contribution of the initial population to the size of the time T population. Aside from contributing to the magnitude of Q, the size and composition of the time zero population is "forgotten", a process known as ergodicity. Equation (2.4) is a statement of the Strong Ergodic Theorem, the "strong" indicating projection by a constant PPM.

In a primitive matrix, $\lambda_1$ is strictly greater than the magnitude $|\lambda_j|$ of any other eigenvalue. The magnitude of positive real eigenvalue $\lambda_1$ is $\lambda_1$. In general, the magnitude of an eigenvalue $\lambda_j = \alpha + \beta i$ (where $\alpha$ and $\beta$ are real and $i$ is the square root of minus one), is

$$|\lambda_j| = [\alpha^2 + \beta^2]^{1/2} \tag{2.5}$$

where the positive square root is taken. In an imprimitive matrix, $\lambda_1 \geq |\lambda_j|$, $j > 1$. Among the subordinate eigenvalues, there are $d - 1$ that have magnitudes equal to $\lambda_1$, even though $\lambda_1$ is greater than the real part of those roots. Such subordinate eigenvalues usually come in complex conjugate pairs. In birth-death models, imprimitivity typically arises when there is only a single reproductive age group. A simple example is a PPM with $\lambda_1 = 1$, but with subordinate eigenvalues equal to the two complex cube roots of 1, i.e. $[-\frac{1}{2} \pm \frac{1}{2}(3)^{\frac{1}{2}}i]$. Over a cycle of length d (here 3), the growth produced by those subordinate eigenvalues equals the growth produced by $\lambda_1$. If the initial population has a state distribution proportional to the dominant right eigenvector of imprimitive PPM **A**, it stays with that distribution. Otherwise, it does not converge to any fixed distribution. Instead, the state composition fluctuates with cycle length d, and the initial state distribution is not forgotten. On average, however, the cyclic population grows by factor $\lambda_1$.

Both primitive and imprimitive matrices are irreducible, that is every state can contribute to every other state. In reducible matrices that is not the case, as at least one state cannot contribute to some other state(s). The crucial criterion is that a $\gamma \times \gamma$ PPM **A** is irreducible if and only if $(\mathbf{I}+\mathbf{A})^{\gamma-1}$ is a positive matrix. Reducible matrices can arise in practice in several ways. One is non-overlapping generations. A second is if post-reproductive age groups are included in a Leslie PPM, as those age groups cannot contribute to any younger age group. A third, which is quite common in multistate analyses, occurs when states are hierarchical, for example a two living state model where there are movements from state 1 to state 2, but not from state 2 to state 1. In a reducible matrix, $\lambda_1 \geq |\lambda_j|$, $\lambda_1$ can equal zero, the dominant right eigenvector may have zero elements, and the initial state distribution of the population may not be forgotten. Sometimes these issues are not problematic, e.g. the lack of a contribution from post-reproductive age groups in birth-death models (Caswell 2001; 88–92) and reducible matrices may yield useful demographic models.

## 2.3 THE SPEED OF CONVERGENCE

Unless otherwise noted, we now consider only primitive PPMs. The trajectory to convergence is provided by

$$\mathbf{x_t} = \mathbf{A^t x_0} \tag{2.6}$$

At every point in that trajectory, the distance to stability and the speed of convergence can be quantified. That is best done using a measure, initially derived in information theory, known as the Kullback distance. Drawing on Tuljapurkar (1982), who brought that measure into the demographic literature, and on Schlögl (1976), we follow Schoen and Kim (1991) in relating the Kullback distance to the process of convergence.

Recall $Q = (\mathbf{v'x_0})$ from equation (2.4). The scalar Q has been termed the stable equivalent, and in birth-death models, the stable equivalent number of births. If the ultimate stable composition was in place at initial time 0, having Q persons age 0 would yield an ultimate stable population with the same size as the one that arises from projecting an actual initial population forward. For any time t, let us adopt continuous notation, focus on birth-death models, and define the stable equivalent by

$$Q(t) = \int N(x, t)\ v(x)\ dx \qquad (2.7)$$

where N(x,t) is the number of persons age x at time t and v(x) is the constant reproductive contribution associated with age x. Reproductive function v(x) gives the contribution from age x to the number of births (or persons in the first age group) in the stable population. Specifically

$$v(x) = \int_x^\infty \exp(-ra)\ p(a)\ f(a)\ da/[A^* \exp(-rx)\ p(x)] \qquad (2.8)$$

where $A^*$ is the mean age at childbearing in the stable population. The integral in equation (2.8) resembles the characteristic equation, but ranges from age x to the end of childbearing. When divided by exp(−rx) p(x), the number in the stable population at age x, it is a reproductive value that represents the "present value" of future children, that is the number of children borne after age x discounted back to age x at "interest rate" r. Division by $A^*$ expresses that reproductive value on a per year basis. In R.A. Fisher's metaphor, stable population characteristic equation (1.29) indicates that the "loan" of a life is "repaid" by net maternity over the life course, including interest that accrues at rate r. In that metaphor, the reproductive value, $v(x)A^*$, represents the amount still owed at age x.

The proportional contribution of the population age x at time t to stable equivalent Q(t) can be designated q(x,t) and written

$$q(x, t) = N(x, t)\ v(x)/Q(t) \qquad (2.9)$$

Since $\int q(x, t)dx = 1$, $q(x, t)$ is a density function. The corresponding reproductive value density distribution in the ultimate stable population, s(x), can be written

$$s(x) = \exp[-rx]\ p(x)\ v(x) \qquad (2.10)$$

The Kullback distance at time t, K(t), measures the distance between the q(x,t) and s(x) distributions. It is defined as

$$K(t) = \int q(x, t) \ln[q(x, t)/s(x)]\ dx \qquad (2.11)$$

Let us specify the "stable equivalent population" at age x and time t, S(x,t), by

$$S(x, t) = Q(t) \exp[-rx]\, p(x) \qquad (2.12)$$

Over time, the stable equivalent population becomes the ultimate stable population. Using equation (2.12), the Kullback distance can be rewritten as

$$K(t) = -\int q(x, t) \ln[S(x, t)/N(x, t)]\, dx \qquad (2.13)$$

where the minus sign preserves a nonnegative K(t). In the form of equation (2.13), the Kullback distance is an interpretable demographic quantity. The ratio S(x,t)/N(x,t) compares the sizes of the stable equivalent population and the observed population at age x and time t. Taking the natural logarithm yields a measure of the amount of growth needed for the observed population to equal its stable equivalent. Accumulating that growth, weighted by each age's contribution to the stable equivalent number of births, provides an overall measure of the growth needed for an observed population to achieve stability.

As the population is projected ahead, K(t) declines monotonically to zero. The size of the decline in K(t) reflects the amount of convergence. Following Schoen and Kim (1991), let us define the force of convergence to stability at time t, $h(t) > 0$, by the log derivative

$$h(t) = -[dK(t)/dt]/K(t) \qquad (2.14)$$

At any time t, K(t) and h(t) provide precise measures of the distance to stability and the speed of convergence to stability, respectively.

It is worth considering whether h(t) approaches a limit as convergence proceeds. Kim and Schoen (1993a) considered that question in some depth, and found that in general the answer is no. If $\lambda_2$ is real, however, there is an constant ultimate force of convergence, $h^*$, given by

$$h^* = 1 - \{|\lambda_2|/\lambda_1\}^2 \qquad (2.15)$$

That relationship does not hold if $\lambda_2$ is complex; in that case the long term force of convergence fluctuates with undiminishing amplitude around the ultimate level given by equation (2.15). In both cases, the long term speed of convergence is negatively related to the *square* of the relative magnitudes of the dominant and second largest eigenvalues of the PPM.

In birth-death models, the net maternity function p(x)f(x) has been parameterized in three principal ways: a normal curve (Lotka), a Type III Gamma function (Wicksell), and an exponential function (Hadwiger). Keyfitz (1977: 156–57) describes those parameterizations and relates them to the moments of the *stable* net maternity function, $e^{-rx}p(x)f(x)$. Using those relationships, Kim and Schoen (1993a) found that $h^*$ can be closely approximated by

$$h^* \approx 1 - \exp[-4n\pi^2\sigma^2/\mu^3] \qquad (2.16)$$

where $\mu$ and $\sigma^2$ are the mean and variance of stable net maternity, n is the length of the projection interval, and $\pi$ is the familiar constant, approximately 3.14159. Equation (2.16) is exact when stable net maternity follows a Hadwiger distribution. The Coefficient of Variation (CV) is defined as the standard deviation divided by the mean, i.e. as $\sigma/\mu$. Equation (2.16) indicates that the long term speed of convergence is directly proportional to the square of the CV and inversely proportional to the mean of stable net maternity. Thus the greater the variability in stable net maternity, and the smaller the stable mean age at childbearing, the faster the pace of convergence. The level of net maternity, i.e. the NRR, is not a factor.

Kim and Schoen (1993a) examined empirical data to evaluate the range of values of $h^*$ and the accuracy of equation (2.16). Using data from Keyfitz and Flieger (1968, 1971) they compared $h^*$ values calculated from equation (2.16) with values from the eigenvalue relationship in equation (2.15). The 2 values were highly correlated. Using 177 populations from Keyfitz and Flieger (1968), $R^2$ was .96, and with 90 populations from Keyfitz and Flieger (1971), $R^2$ was .98. Knowledge of the first two moments of stable net maternity thus allows $h^*$ to be found with considerable accuracy.

Entropy, a measure of the randomness (or information content) in a distribution is an important concept in physics that plays a significant role in demographic analyses. If q represents a density distribution, its entropy has the basic form $\int q \ln q$, though the concept has been specified in a number of ways in different applications. Entropy thus has a form similar to that of the Kullback distance. Tuljapurkar(1982;1993) discussed relationships between entropy and convergence, and showed that entropy provides a lower bound on the rate of convergence. To compare entropy and $h^*$, we follow Kim and Schoen (1993a), assume stable net maternity can be described by a normal curve, and write entropy (H) in terms of the moments of stable net maternity as

$$H = \{\ln[\sigma(2\pi)^{1/2}] + 1/2\}/\mu \qquad (2.17)$$

Essentially, entropy varies with $\ln[\sigma]/\mu$, while $h^*$ varies with $[\sigma^2/\mu^3]$. The two are not identical, but they move in the same direction and both reflect the underlying dispersion of stable net maternity. In a regression analysis based on 177 populations in Keyfitz and Flieger (1968), Kim and Schoen (1993a) found the linear relationship

$$H = .2089 + .2620\,h^* \qquad (2.18)$$

with $R^2 = .95$.

*Table 2.1. Demographic Measures Related to Convergence for 3 Selected Female Populations*

| | Measure | Population | | |
|---|---|---|---|---|
| | | Japan 1963 | Togo 1961 | United States 1959–61 |
| 1. | Expectation of Life at Birth | 72.33 | 40.12 | 66.84 |
| 2. | Net Reproduction Rate | 0.926 | 2.141 | 1.713 |
| 3. | Mean Age of Stable Net Maternity | 27.85 | 27.06 | 25.51 |
| 4. | Variance of Stable Net Maternity | 17.09 | 50.48 | 31.76 |
| 5. | Entropy (H) | .2539 | .3323 | .3182 |
| 6. | Ultimate Level of the Force of Convergence ($h^*$) | .2062 | .5019 | .4096 |
| 7. | Kullback Distance ($\times 1000$), Year 0 | 5.97724 | 45.85946 | 3.30507 |
| 8. | Kullback Distance ($\times 1000$), Year 100 | 0.06011 | 0.00007 | 0.00009 |

*Source:* Kim and Schoen (1993), with data from Keyfitz and Flieger (1968). Reprinted with permission from *Mathematical Population Studies*, Taylor and Francis Group, LLC, http://taylorandfrancis.com, (c) 1993.

To examine empirical patterns of convergence, Kim and Schoen (1993a) used data from Keyfitz and Flieger (1968) for three populations with quite different patterns of fertility and mortality. Table 2.1 shows selected values for Japan 1963, Togo 1961, and the United States 1959–61. The level of $h^*$ varied from around 0.2 to 0.5, indicating fairly rapid convergence. The long term fluctuations around $h^*$ were modest for the United States and even less substantial for Togo, but were quite pronounced for Japan. Keyfitz (1972) found that the second largest root is responsible for most of the fluctuations that occur during the transition to stability; effects from the other subordinate roots attenuate very quickly. The second largest root typically has a cycle length that approximates the length of a generation (Keyfitz 1977: Ch. 6), and under the Hadwiger distribution that cycle length exactly equals $A^*$, the stable mean age at childbearing.

## 2.4 SOME EXPLICIT ANALYSES OF CONVERGENCE

We consider the process of convergence to stability in some detail in the context of three relatively simple models. The first is a birth-death model with

2 ages, the second a birth-death model with 3 ages, and the third a 2 living state multistate model without age.

### 2.4.1 Convergence in 2 × 2 Leslie Models

In a 2 age group Leslie model, the discrete characteristic equation can be written

$$1 = b_1/\lambda + s_1 b_2/\lambda^2 = a/\lambda + b/\lambda^2 \qquad (2.19)$$

where the $s_i$ and $b_j$ follow the notation used in equation (1.41), with $a = b_1$ and $b = s_1 b_2$ introduced as simplifications. The NRR is $(a+b)$. Equation (2.19) is a quadratic in $\lambda$ whose 2 roots are

$$\lambda = \tfrac{1}{2}[a \pm (a^2 + 4b)^{1/2}] \qquad (2.20)$$

where $\lambda_1$ is associated with the positive root and both roots are real. From the relationship between the roots of a quadratic and coefficients a and b (see Exercise 1), we have

$$h^* = 1 - (1 - a/\lambda_1)^2 \qquad (2.21)$$

The more net maternity is concentrated in the first age group, the faster the ultimate rate of convergence.

### 2.4.2 Convergence in a 3 × 3 Leslie Model

The discrete characteristic equation of a 3 age group Leslie model is given by

$$\begin{aligned} 1 &= b_1/\lambda + s_1 b_2/\lambda^2 + s_1 s_2 b_3/\lambda^3 \\ &= a/\lambda + b/\lambda^2 + c/\lambda^3 \end{aligned} \qquad (2.22)$$

where $a = b_1$, $b = s_1 b_2$, $c = s_1 s_2 b_3$, and the NRR is $(a+b+c)$. Equation (2.22) is cubic, and as usual we let its dominant root be $\lambda_1$. To simplify the solution, we use the transformed values $a' = a/\lambda_1$, $b' = b/\lambda_1^2$, and $c' = c/\lambda_1^3$. Here a', b', and c' are the elements of the discrete stable net maternity function, and they sum to 1. With $\lambda' = \lambda/\lambda_1$, equation (2.22) can be rewritten as the quadratic

$$(\lambda')^2 + (1 - a')\lambda' + c' = 0 \qquad (2.23)$$

(see Exercise 1). The roots of equation (2.23) are

$$\{\lambda_2', \lambda_3'\} = \tfrac{1}{2}[-(1 - a') \pm \{(1 - a')^2 - 4c'\}^{1/2}] \qquad (2.24)$$

with $\lambda_1' = 1$ and $\lambda_2'$ assigned to the positive root of equation (2.24). Subordinate roots $\lambda_2'$ and $\lambda_3'$ can be either real and negative or complex conjugates. Reasonable values of net maternity can lead to either pattern of roots.

If the discriminant, the term in curly brackets in equation (2.24) whose square root is taken, is nonnegative, the roots are real, and

$$c' \leq (1 - a')^2/4$$

In this case, equation (2.24) implies that the larger the value of c', the stable net maternity value for the highest age group, the smaller the magnitude of $\lambda_2'$ and the larger the value of h*. If the discriminant is negative, there are 2 complex conjugate roots. Equation (2.24) then gives $|\lambda_2'| = [c']^{1/2}$, which also implies $h^* = 1 - c'$. When the subordinate eigenvalues are complex, the larger the value of c' the larger the magnitude of $\lambda_2'$ and the smaller the long term speed of convergence.

To summarize the relationship between c' and h*, we begin with c' small and both subordinate roots real. As c' increases, $|\lambda_2'|$ decreases and h* becomes larger, up to the point where the discriminant equals zero. At that point, the subordinate roots are real and equal, and h* attains its maximum (for that value of a'; the larger a', the greater the maximum). As c' continues to increase, the subordinate roots become complex and h* declines, accompanied by oscillations of increasing amplitude.

The Leslie model with three reproductive age groups, ages 0–14, 15–29, and 30–44, captures much of the dynamics of models with 5-year age groups, but is more amenable to explicit algebraic analysis. That is especially true when there is a double root (i.e. $\lambda_2 = \lambda_3$). While multiple roots usually complicate matters, here they reduce a cubic equation to a quadratic and allow an explicit expression for the birth trajectory to stability (cf. Schoen and Kim 1996).

The eigenvalues in the double root case are dominant root $\lambda = \lambda_1$ and double root $\lambda_2 = -\lambda[c']^{1/2}$, with $a' = 1 - 2[c']^{1/2}$ and $b' = 2[c']^{1/2} - c'$. Here PPM **A** does not have a full complement of eigenvectors, but can be put in Jordan canonical form (Impagliazzo, 1985; Noble and Daniel, 1988, see Exercise 7). Raising **A** to successive powers yields the birth trajectory, $B_t$, i.e. the number of persons in the first age group at time t. With births discounted for stable growth,

$$B_t/\lambda^t = Q + \{-[c']^{1/2}\}^t[(1 - Q) + t\{1 + \lambda[c']^{1/2}x_{20}/s_1 - Q(1 + [c']^{1/2})\}] \qquad (2.25)$$

where $x_{j0}$ represents the number of persons in age group j at initial time 0. The stable equivalent number of births in equation (2.25) is given by

$$Q = \{1 + 2\lambda[c']^{1/2} x_{20}/s_1 + \lambda^2 c' x_{30}/(s_1 s_2)\}/\{1 + [c']^{1/2}\}^2 \qquad (2.26)$$

In the long term, $B_t = Q\lambda^t$, as indicated by equation (2.3). In the short term, there are fluctuations around that exponential trajectory on the order of $t\{-[c']^{1/2}\}^t$, which alternate in sign and diminish by a factor of approximately $[c']^{1/2}$ every 15 years. When the initial population is stable with growth parameter 1.25, Figure 2.1 shows the trajectory to stationarity when $\lambda = 1$, for the double root case (c'=0.1225) and for neighboring values of c'. The birth trajectories are very similar for all values of c', and show initially sizeable but rapidly dampening fluctuations in the number of births. Larger values of c' are associated with a smaller ultimate number of births. In most empirical populations, after 60 years

*Figure 2.1. Birth Trajectories to Stationarity in the Neighborhood of a Double Root, Leslie Model with 3 Age Groups*

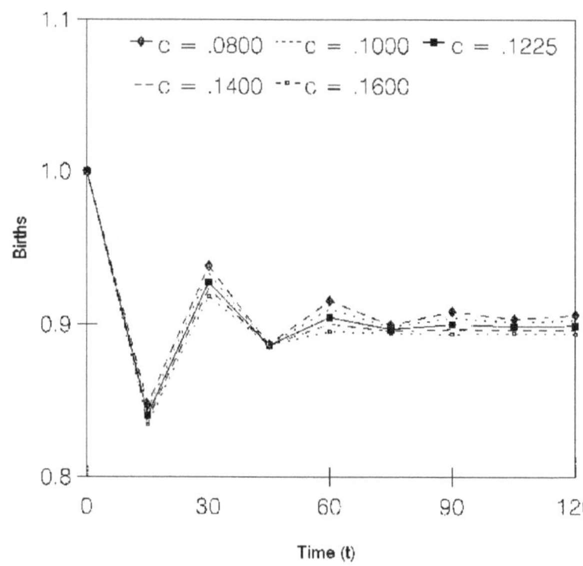


*Calculated by projection from an initial population given by $x_0 = (1, (1.25)^{-1}, (1.25)^{-2})'$. The double root net maternity pattern is (.3, .5775, .1225). In all cases, $a = .3$.*

*Source:* Schoen and Kim (1996, Figure1b). Reprinted with permission from *Mathematical Population Studies*, Taylor and Francis Group, LLC, http:// taylorandfrancis.com, (c) 1996.

the size of the birth cohort is typically close to its ultimate stationary level, and the total size of the population is also close to its final size. As a rough rule of thumb, most initial populations approximate their ultimate stationary values up to age x after 60 + x years.

Birth fluctuations produced by a rapid decline in fertility can substantially impact the age composition of a stabilizing population. Sixty years after the shift to a stationary net maternity regime, the largest (time 0) birth cohort is at ages 60–74, while the smallest cohort is at ages 45–59. That can produce a "surge" in the proportion of the population over age 60 that goes beyond the long term population aging associated with fertility declines (a topic discussed in Chapter 3). Schoen (1996) found that the surge was closely approximated by 1/Q. With the initial population stable with growth parameter $\zeta$,

$$Q \approx (\zeta + 1)/(2\zeta) \tag{2.27}$$

If $\zeta = 1.25$ (i.e. Lotka's $r = 0.0149$), then $Q \approx 0.90$ and the surge, $1/.9 = 1.11$, increases the proportion of the population over age 60 at Year 60 by a factor of 11% above its ultimate stationary value.

Earlier in the transition, compositional changes can have the effect of temporarily reducing a population's dependency burden. Let the total dependency ratio be the population under age 15 and over age 65 divided by the population at ages 15–64. After a drop in fertility, the population under age 15 falls, but it is some time before the number over age 65 rises appreciably. In the intervening period, which can last several decades, the dependency ratio can fall below 0.5. That "demographic window" or "demographic bonus" can provide an opportunity for accelerating economic development (Birdsall and Sinding 2001; Robine, Cheung, Tu and Zeng 2003).

### 2.4.3 Convergence in a Two Living State Multistate Model

Consider a multistate population with two living states, A and B, governed by the differential equations

$$\begin{aligned} dA(t)/dt &= A(t)[\phi_A - \mu_{AB} - \mu_{A\delta}] + B(t)\mu_{BA} \\ dB(t)/dt &= B(t)[\phi_B - \mu_{BA} - \mu_{B\delta}] + A(t)\mu_{AB} \end{aligned} \tag{2.28}$$

where $\phi$ indicates the fixed force of fertility and $\mu_{IH}$ is the constant force of decrement from state I to state H, with $\delta$ indicating the "dead" state. All of the forces are non-negative, and either $\mu_{AB}$ or $\mu_{BA}$ is positive to insure that the two states intercommunicate. We want to examine how such a model moves from arbitrary initial composition A(0) and B(0) to the stable state composition.

It is useful to consider the "autonomous" growth rate of each state, that is its rate of natural increase in the absence of interstate movements. Those autonomous growth rates are

$$\begin{aligned} r_A &= \phi_A - \mu_{A\delta} \\ r_B &= \phi_B - \mu_{B\delta} \end{aligned} \tag{2.29}$$

It is also convenient to introduce the symbols

$$\begin{aligned} \alpha &= r_A - \mu_{AB} \\ \beta &= r_B - \mu_{BA} \end{aligned} \tag{2.30}$$

Differential equations (2.28) can be written in matrix form as

$$d\mathbf{x(t)}/dt = \boldsymbol{\mu}\ \mathbf{x(t)} \tag{2.31}$$

where $\mathbf{x(t)}$ is a column vector with elements A(t) and B(t) and $\boldsymbol{\mu}$ is the matrix of forces (or rates)

$$\boldsymbol{\mu} = \begin{bmatrix} \alpha & \mu_{BA} \\ \mu_{AB} & \beta \end{bmatrix} \tag{2.32}$$

The eigenvalues of $\boldsymbol{\mu}$, dominant root r and subordinate root s, are given by

$$r, s = \tfrac{1}{2}\{(\alpha+\beta) \pm [(\alpha-\beta)^2 + 4\mu_{AB}\mu_{BA}]^{1/2}\} \tag{2.33}$$

which satisfies the trace relationship $(r+s) = (\alpha+\beta)$. The right eigenvector matrix of $\boldsymbol{\mu}$ can be written

$$\mathbf{U} = \begin{bmatrix} 1 & 1 \\ \mu_{AB}/(r-\beta) & \mu_{AB}/(s-\beta) \end{bmatrix} = \begin{bmatrix} 1 & 1 \\ (r-\alpha)/\mu_{BA} & (s-\alpha)/\mu_{BA} \end{bmatrix} \tag{2.34}$$

which implies the relationship

$$(r-\alpha)(r-\beta) = \mu_{AB}\mu_{BA} = (s-\alpha)(s-\beta) \tag{2.35}$$

If either $\mu_{AB}$ or $\mu_{BA}$ is zero, r must be equal to either $\alpha$ or $\beta$. If $\mu_{AB} = 0$, we must have $r = \beta$, and hence $\beta > \alpha$, or the proportion in state B goes to zero. With one transfer rate equal to zero, the model's PPM is reducible (see Exercise 4).

The reproductive contributions matrix, $\mathbf{V} = \mathbf{U}^{-1}$, can be written

$$\mathbf{V} = \begin{bmatrix} (r-\beta) & -(r-\beta)(s-\beta)/\mu_{AB} \\ -(s-\beta) & (r-\beta)(s-\beta)/\mu_{AB} \end{bmatrix} \{1/(r-s)\} \tag{2.36}$$

The first column of $\mathbf{V}$ sums to 1 and the second column sums to zero. The (1,2) element of $\mathbf{V}$ can also be written $\mu_{BA}/(r-s)$ as $\mu_{AB}\mu_{BA} = -(r-\beta)(s-\beta) = -(r-\alpha)(s-\alpha)$.

The general solution to differential equations (2.28) can be obtained in a straightforward fashion (e.g. Ford 1955), and written (cf. Ledent 1978; Schoen and Kim 1993)

$$A(t) = Q_1 e^{rt} + Q_2 e^{st}$$
$$B(t) = \{\mu_{AB}/(r-\beta)\}\ Q_1 e^{rt} + \{\mu_{AB}/(s-\beta)\}\ Q_2\ e^{st} \tag{2.37}$$

where $Q_1$ and $Q_2$ are stable equivalents, i.e. $Q_1$ is the vector product of the first row of $\mathbf{V}$ in equation (2.36) and $\mathbf{x_0}$, and $Q_2$ is the product of the second row of $\mathbf{V}$ and $\mathbf{x_0}$. Note that the elements in the second row of $\mathbf{U}$ are part of the equation for B(t). In equations (2.37), the $e^{rt}$ terms reflect the number in the dominant component of the population that grows to become the multistate stable population, while the $e^{st}$ terms reflect the number in the subordinate component that declines over time to zero (or insignificance).

An alternative approach to differential equations (2.28) is to solve matrix equation (2.31). Since $\mu$ is constant, that yields

$$\begin{aligned}\mathbf{x(t)} &= \exp(\boldsymbol{\mu}\mathbf{t})\ \mathbf{x(0)}\\ &= \mathbf{U}\exp(\mathbf{R}t)\mathbf{V}\ \mathbf{x(0)}\end{aligned} \tag{2.38}$$

where $\mathbf{R}$ is the eigenvalue matrix of $\boldsymbol{\mu} = \mathbf{URV}$, which has r as its first diagonal element and s as its second diagonal element. The PPM eigenvalue matrix $\Lambda = \exp(\mathbf{R}t)$ is a diagonal matrix whose elements are $e^{rt}$ and $e^{st}$. The population projection matrix $\mathbf{A^t} = \exp(\boldsymbol{\mu}t)$ takes the population from time 0 to time t. Expanding matrix equation (2.38) yields the solution in equations (2.37).

Applying the definition of the Kullback distance, Schoen and Kim (1993) derived a general expression for K(t). At large time T,

$$K(T) = -\exp[-2T(r-s)][Q_2/Q_1][\{Q_2\mu_{AB}/(s-\beta)\}/\{Q_1\mu_{AB}/(r-\beta)\}] \tag{2.39}$$

In words, that Kullback distance reflects the product of (i) the initial ratio of the size of the subordinate component to that of the dominant component in state A times (ii) that same ratio in state B, which is (iii) discounted for T years of growth in each state by the extent that r exceeds s. The Kullback distance is always positive; the minus sign in equation (2.39) reflects that the subordinate component in either state A or state B is negative.

The long term force of convergence, minus the log derivative of K(T), is then

$$h^* = 2(r-s) = 2[(\alpha-\beta)^2 + 4\mu_{AB}\mu_{BA}]^{1/2} \tag{2.40}$$

Since $r > s$, $h^*$ is positive. For a given pair of transfer rates, $h^*$ is minimized when $\alpha = \beta$. When the autonomous growth rates are given, $h^*$ is minimized when $|r_A - r_B| = \mu_{AB} + \mu_{BA}$. Hence increases in the rates of interstate transfer accelerate the speed of convergence whenever their sum is larger than the difference between the autonomous growth rates.

Theoretically $h^*$ has no upper bound, but human growth rates are typically small. In an application to an urban/rural model using hypothetical but plausible data, Schoen and Kim (1993) found $h^*$ values on the order of 0.02. That is smaller than the $h^*$ values found in birth-death models by an order of magnitude, and it implies that convergence to stability in multistate models is much slower than in birth-death models. In one instance, Schoen and Kim (1993) found a 3% difference between the proportion in a state 500 years after time 0 and the ultimate stable proportion. Liaw (1980), in a multistate analysis of the Canadian population by age and region, found that after 100 years the age composition of the projected population was close to its stable values, while the regional distribution was perhaps halfway to stability.

## 2.5 SUMMARY

This chapter examines the process of stabilization through which an initial population becomes stable under a regime of time invariant demographic rates. Convergence to stability is assured only for primitive population projection matrices, where the dominant root is larger in magnitude than any subordinate root. If a $\gamma \times \gamma$ PPM raised to the $\gamma(\gamma - 2) + 2$ power has all positive elements, it is primitive, as is any Leslie matrix with net maternity in adjacent age groups.

The Kullback distance, a measure from information theory, reflects a population's distance to stability and, over time, the pace at which that distance monotonically declines to zero during convergence. The long term speed of convergence depends on the square of the ratio of the magnitudes of the two largest roots, but in general does not approach a fixed limit. In birth-death models, the speed of convergence is greater when net maternity has a larger coefficient of variation and a smaller mean age. Typically, stabilization up to age x is approximately complete after $60 + x$ years.

The process of convergence is more transparent in relatively simple models. Leslie matrices with 2 and 3 age groups are examined, and a complete birth trajectory to stability is presented for a 3 age group model with equal subordinate roots. A complete trajectory is also given for a two living state multistate model, where convergence to the stable state composition is typically much slower.

## 2.6 EXERCISES

[See Appendix B for selected answers]

1. Characteristic equation (2.22) can be rewritten as

$$\lambda^3 - a\lambda^2 - b\lambda - c = 0$$

with roots $\lambda_1$, $\lambda_2$, and $\lambda_3$.

a. Show that $a = \lambda_1 + \lambda_2 + \lambda_3$
$b = -(\lambda_1\lambda_2 + \lambda_1\lambda_3 + \lambda_2\lambda_3)$
$c = \lambda_1\lambda_2\lambda_3$

b. With transformed coefficients a', b', and c' (as defined after equation (2.22)), show that all of the roots are divided by $\lambda_1$.

c. Show that in the quadratic equation

$$\zeta^2 - \alpha\zeta - \beta = 0$$

with roots $\zeta_1$ and $\zeta_2$ that $\alpha = \zeta_1 + \zeta_2$ and $\beta = -(\zeta_1\zeta_2)$

d. Verify equation (2.24)

2. Consider the PPM

$$\begin{bmatrix} 0\,1\,0\,1 \\ 1\,0\,0\,0 \\ 0\,1\,0\,0 \\ 0\,0\,1\,0 \end{bmatrix}$$

a. Determine whether it is primitive, imprimitive, or reducible.

b. Evaluate the matrix raised to the 20th power.

3. Consider the following PPMs

a. $\begin{bmatrix} 0\,1 \\ 1\,0 \end{bmatrix}$

b. $\begin{bmatrix} 0\,0\,1 \\ 1\,0\,0 \\ 0\,1\,0 \end{bmatrix}$

Show that they are imprimitive. Find their cycle lengths, roots, and cyclic patterns.

4. Consider the rate matrix

$$\boldsymbol{\mu} = \begin{bmatrix} \alpha & 0 \\ \mu_{AB} & \beta \end{bmatrix}$$

a. Find the associated population projection matrix **A**. Satisfy yourself that all elements of **A** are nonnegative.

b. Verify that **A** is reducible.

5. Consider a $3 \times 3$ Leslie matrix with first row elements a, b, and c. If $c = \frac{1}{4}(1-a)^2$ and $b = 1 - a - c$, show that the matrix has a double root.
6. Show that if the initial population is $\mathbf{x}_0' = (1, \zeta^{-1}, \zeta^{-2})$, where the $'$ indicates a matrix transposition, then equations (2.25) and (2.26) become

a. $B_t/\lambda^t = Q + \{-[c']^{1/2}\}^t[(1-Q) + t[c'Q]^{1/2}(1-\lambda/\zeta)]$

b. $Q = \{(1+[c']^{1/2}\lambda/\zeta)/(1+[c']^{1/2})\}^2$

7. In the $3 \times 3$ double root case, PPM **A** can be written $\mathbf{A} = \mathbf{WJW}^{-1}$, where **J** is the Jordan form matrix

$$\mathbf{J} = \begin{bmatrix} \lambda & 0 & 0 \\ 0 & \lambda_2 & 1 \\ 0 & 0 & \lambda_2 \end{bmatrix}$$

and **W** is

$$\mathbf{W} = \begin{bmatrix} 1 & 1 & 1 \\ s_1/\lambda & s_1/\lambda_2 & (s_1/\lambda_2)(1-1/\lambda_2) \\ s_1 s_2/\lambda^2 & s_1 s_2/\lambda_2^{\,2} & (s_1 s_2/\lambda_2^{\,2})(1-2/\lambda_2) \end{bmatrix}$$

Show that

a. $\mathbf{A}^t = \mathbf{WJ}^t\mathbf{W}^{-1}$

b. $\mathbf{J}^t$ has the form

$$\mathbf{J}^t = \begin{bmatrix} \lambda^t & 0 & 0 \\ 0 & \lambda_2^{\,t} & t\,\lambda_2^{\,t-1} \\ 0 & 0 & \lambda_2^t \end{bmatrix}$$

8. With $\lambda_1 = e^r$ and $\lambda_2 = e^s$, show that, to linear terms, the $h^*$ of equation (2.15) is the same as the $h^*$ of equation (2.40).
9. Espenshade, Bouvier, and Arthur (1982) analyzed a population with constant, below replacement, vital rates that received a constant number of inmigrants, by age, each year. Assume that, in the long term, the migrants have $B_1$ female births each year. Verify Espenshade et al's finding that every year the total number of births in that ultimate stationary population is $B = B_1/(1 - NRR)$.
10. Describe how an arbitrary population, with 15 year age intervals, can achieve a given stable composition through age 45 in 45 years.

# CHAPTER 3

# POPULATION MOMENTUM

## 3.1 INTRODUCTION

Population momentum is the long term change in size that results from a change in vital rates, most frequently to a regime that yields an ultimately stationary population. As discussed in Chapter 2, after convergence, a population subject to a fixed set of demographic rates has a state composition and rate of growth that depends only on those fixed rates. The population's previous size and state composition are forgotten, except for an enduring influence on the ultimate population's size. In this chapter, we examine the concept of population momentum, its importance in demographic theory and practice, how it can be measured, and how it relates to other demographic quantities. After considering birth-death models at some length, we turn to multistate models and explore the phenomenon of spatial momentum.

## 3.2 MOMENTUM AS A DEMOGRAPHIC PHENOMENON

### 3.2.1 The Basic Dynamics

The analysis of population momentum dates from Keyfitz (1971), which examined a growing stable population that suddenly experiences a fall in fertility to replacement level (i.e. an NRR of 1), maintaining those replacement level rates thereafter. Momentum is defined as the ratio of the ultimate stationary population to that of the initial population. With mortality remaining constant and the decline in fertility proportional at all ages, Keyfitz (1971) showed that momentum, $\Omega$, can be expressed as

$$\Omega = \mathrm{b}\ \mathrm{e}(0)\ \mathrm{Q} \qquad (3.1)$$

where b is the intrinsic birth rate of the stable population and e(0) the constant life expectancy at birth. Recall that in equation (1.33) b is the reciprocal of the size of the initial stable population with a unit birth at the time of the fertility decline, and that e(0) is the size of the stationary population given a unit birth each year. Thus the product be(0) represents the ratio of the size of the ultimate stationary population to that of the initial stable population, with both having birth cohorts of the same size. However, given a unit birth for the initial stable population, the size of every birth cohort in the ultimate stationary population is not 1 but Q, the stable (here stationary) equivalent number of births. In the scenario of Keyfitz (1971),

$$Q = (R_0 - 1)/(r\ \mu\ R_0) \tag{3.2}$$

where $R_0$ is the stable population Net Reproduction Rate, r is the intrinsic growth rate, and $\mu$ is the mean age at net maternity in the stationary population.

It is not realistic to assume that the fertility of a rapidly growing population will immediately drop to replacement level, but that assumption can provide a *minimum* estimate for future population growth. Keyfitz (1971) examined some contemporary populations and found a number of momentum values around 1.5 to 1.7, indicating that national population growth of some 50% to 70% could result from population momentum alone. The importance of momentum as a major source of population growth was reinforced in a recent comprehensive analysis by Bongaarts and Bulatao (1999), which showed that population momentum is likely to be responsible for most of the future growth of the world's population.

To examine the characteristics of the growth associated with momentum, let us look at equation (3.1) more closely. When $r > 0$, $R_0 > 1$ and $Q < 1$. Because fertility falls when the initial $R_0$ is above replacement, one should expect that the stationary population birth cohort would be smaller than the initial stable birth cohort. To a reasonable approximation (Keyfitz 1985:157; see Exercise 1)

$$Q \approx (R_0)^{-1/2} \tag{3.3}$$

Thus if the initial population $R_0 = 1.69$ then $Q \approx 1/1.3$ or about 0.769.

Since birth cohort size decreases, the growth in population must come from the ratio be(0). With initial and ultimate birth cohorts the same size, we can use equation (1.32) to write

$$b\ e(0) = \int \ell\ (a)\ da / \int \exp(-ra)\ell(a)\ da \tag{3.4}$$

where as usual the integrals range from 0 to $\infty$ and $\ell(a)$ represents the number of survivors to exact age a from a cohort of $\ell(0)$ births. With $r > 0$, the numbers at the higher ages in the stable population are substantially reduced in size, relative to those in the life table population, because they are the survivors of smaller birth cohorts, whose size is discounted for many years of exponential growth at rate r. The further back in time, the greater the differential cohort size, hence the larger the value of r, the greater the population momentum. For example, in a female stable population with $e(0) = 75$, population size is 75 when $r = 0$, 52.03 when $r = .01$, and 38.01 when $r = .02$ (see Coale and Demeny 1966:70). Thus $be(0)$ is 1.44 when $r = .01$ and 1.97 when $r = .02$. In contrast, Q is about 0.87 when r is .01 and 0.75 when $r = .02$, yielding momentum values of approximately 1.25 and 1.48, respectively.

With sustained below replacement fertility in a number of countries, more attention is being given to cases of "negative momentum", i.e. populations whose rates imply $r < 0$ and where $NRR < 1$ (e.g. Lutz, O'Neill and Scherbov 2003). The same equations apply in both cases. When net reproduction is less than 1, $Q > 1$, $be(0) < 1$, and $\Omega < 1$. Although the ultimate stationary population has a larger birth cohort size, it has smaller numbers at the higher ages and thus becomes smaller over the transition to stationarity. As the momentum associated with past population increase leads to more growth, the momentum associated with past population decrease leads to more decline.

Equation (3.1) can readily be modified to apply to the case where any initial population has an arbitrary change in vital rates to replacement level, with those new rates continuing unchanged. Specifically we can write

$$\Omega = BR\, e(0)\, Q \qquad (3.5)$$

where BR is the initial birth rate, e(0) is the ultimate stationary population life expectancy, and the stable equivalent number of births, $Q = \mathbf{v}'\mathbf{x}_0$, is based on the ultimate stationary rates and the initial population (see equations 2.4, 2.7, and 2.8).

The concept of momentum can be extended even further, to the size adjustment implied by any change in vital rates. In examining the Kullback distance, Schoen and Kim (1991) defined an age x, time t specific momentum, $\Omega(x, t)$, as

$$\Omega(x, t) = S(x, t)/N(x, t) \qquad (3.6)$$

where S(x,t) is the stable equivalent population age x at time t under the prevailing rates [see equation 2.12] and N(x,t) is the population age x at time t. Under that definition, the prevailing rates need not lead to a stationary population, and overall momentum is just the ratio of the total stable equivalent population at time t to the total time t population. In certain circumstances it is useful to think of momentum in those terms, but there are two caveats. First, the size adjustment may be purely theoretical, as the rates may continue to change. Second, that generalized momentum ignores the future stable growth that occurs when the rates do not imply long term stationarity. The stable equivalent size adjustment from the shift in rates may be dwarfed by differences in long term growth.

### 3.2.2 Momentum and Aging

Preston (1986), an insightful paper, considered the age compositional implications of the growth associated with conventional population momentum. It argued that up to an age approximating the length of a generation, the initial and ultimate stationary populations have the same *number* of persons. Thus all of the growth associated with momentum comes above that age. Heuristically, Preston (1986) divided an initial population into 3 generation-long segments, intervals of some 28 to 30 years each. In the transition from initial conditions to the ultimate stationary population, the size of the first segment remains unchanged; the size of the second segment increases by a factor approximating the initial NRR, i.e. the amount of growth over a generation; and the size of the third segment increases by a factor approximating two generation's growth, i.e. the NRR squared.

Preston (1986) took Lotka's generation length, Ŧ, as the age up to which there was no change in size over the transition to stationarity. Further research, however, showed that Lotka's Ŧ does not always provide a good estimate (Wachter, 1988; Kim and Schoen, 1993b, see Section 4.4). Kim, Schoen and Sarma (1991) noted that momentum could be expressed as the ratio of initial to ultimate population proportions through age G, the age up to which there is no change in size, or

$$\Omega \approx C_N(0, G)/C_L(0, G) \tag{3.7}$$

where C(0,G) represents the fraction of the population between ages 0 and G, subscript N indicates the initial population, and subscript L indicates the ultimate stationary (life table) population. To determine age G, Kim et al (1991) suggested the geometric mean relationship

$$G \approx [\mu A_N]^{1/2}$$

where $\mu$ and $A_N$ are the stationary and initial population mean ages at child-bearing, respectively.

Figure 3.1, adapted from that work, shows population sizes up to each age based on the female birth and death rates observed in Singapore 1957. The size of the (implicit) initial stable population has been rescaled to 1. The total size of the ultimate stationary population, 1.75, is thus the stable population's momentum based on that implicit initial population. Below the age where the two curves cross, about age 30, there are more people in the initial population; after that age, there are more in the ultimate population, and the gap widens with age.

*Figure 3.1. Cumulative Initial and Ultimate Populations, By Age, Based on Rates Observed in Singapore 1957*

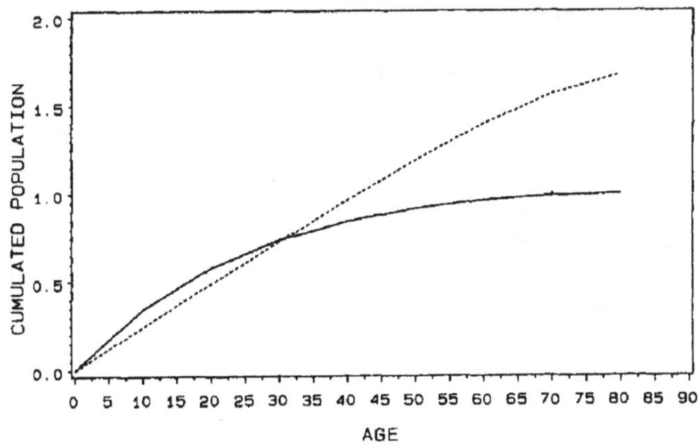


*Note:* Number of persons up to each age in the initial stable population (solid line), and the ultimate stationary population (broken line). The cumulative (total) size of the initial stable population has been scaled to 1. The cumulative size of the ultimate stationary population, $\Omega$, is 1.75.
*Source:* Kim, Schoen and Sarma (1991, Figure 2).

Kim and Schoen (1997) further explored the relationship between momentum and population aging. Using stable population relationships, momentum could be approximated by

$$\Omega \approx 1 + r\ (Y_L - V_L) \qquad (3.8)$$

where $Y_L$ is the mean age of the ultimate stationary population and $V_L$ is the stationary population mean age of reproductivity. That mean of the stationary population reproductive value distribution is not well known, but is a bit over half the mean age at net maternity and, to second order terms, is given by

$$V_L = \mu(1 + [\sigma^2/\mu^2])/2 \qquad (3.9)$$

where $\mu$ and $\sigma^2$ are the mean and variance of net maternity in the stationary population. The initial stable population mean age, $Y_r$, can be related to the ultimate stationary population mean age by the approximation

$$Y_r \approx Y_L - r\ \sigma^2{}_Y \qquad (3.10)$$

where $\sigma^2{}_Y$ is the variance of the stationary population age distribution. Combining equations (3.8) and (3.10) gives an approximate linear relationship between momentum and aging, that is

$$\Omega \approx 1 + \Delta A[(Y_L - V_L)/\sigma^2{}_Y] \qquad (3.11)$$

where $\Delta A = Y_L - Y_r$ is the increase in the mean age of the population over the transition from stable to stationary.

That linear relationship between momentum and aging holds up extremely well under both theoretical and empirical analysis. Using 39 Coale and Demeny (1966) Model West female stable populations, with e(0) values of 60, 70, and 80 and r values ranging from .01 to .05, Kim and Schoen (1997) estimated the regression equation

$$\begin{array}{c} \Omega = .997 + .047\ \Delta A \\ (.007)\ (.001) \end{array} \qquad (3.12)$$

where the standard errors are shown in parentheses and the adjusted $R^2$ is .995. The model population results in equation (3.12) are consistent with the intercept of 1 predicted by equation (3.11), and close to the predicted slope, as the model population values of $(Y_L - V_L)/\sigma^2{}_Y$ ranged from .041 to .044. With no change in the mean age of the population, $\Omega = 1$. Each year of increase in mean population age is associated with population growth of 4.7%.

Kim and Schoen (1997) estimated the same model using the 41 female populations for the year 1985 given in Keyfitz and Flieger (1990). The regression equation is

$$\begin{array}{c} \Omega = .980 + .047\ \Delta A \\ (.007)\ (.001) \end{array} \qquad (3.13)$$

very close to the previous result and with an adjusted $R^2$ of .985. Such a close fit with empirical data is remarkable, demonstrating the strength of the linear relationship between momentum and the increase in mean population age.

Figure 3.2, taken from Kim et al (1991), shows a striking constancy in the number of persons under age 30 during the transition to stationarity of the observed population of Singapore 1957. Kim and Schoen (1997) used the same approach and data sets to examine the appropriateness of using age 30 as the age under which the numbers in the initial and ultimate populations are identical. Let $F_{30}$ be the ratio of the initial population proportion under age 30 to the ultimate stationary population proportion under age 30, i.e.

$$F_{30} = C_N(0, 30)/C_L(0, 30) \qquad (3.14)$$

*Figure 3.2. Projection to Stationarity of the Observed Population of Singapore 1957*

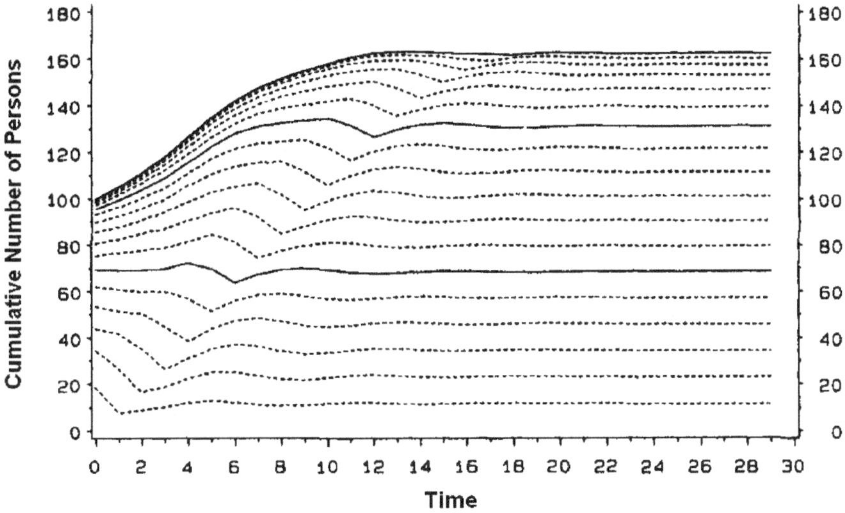


*Note:* Figure shows the evolution of population segments from age 0 to age 90, in 5-year age groups. Solid lines represent ages 30, 60, and 90. Time is in units of 5 years. The initial size of the population was rescaled to 100, and the ultimate size of the projected population is 162.
*Source:* Kim, Schoen and Sarma (1991, Figure 1).

The expected relationship is simply $\Omega = F_{30}$. The model population regression yielded

$$\Omega = .003 + .998\ F_{30} \tag{3.15}$$

with an adjusted $R^2$ of .9997, and the empirical population regression gave

$$\Omega = .003 + .989\ F_{30} \tag{3.16}$$

with an $R^2$ of .997. Momentum was virtually identical to the ratio of the initial population fraction under age 30 to the ultimate population fraction under age 30.

In sum, demographic momentum is essentially the long term change in the number of persons over age 30 that occurs when an initial population abruptly shifts to replacement level vital rates. Theoretically and empirically, the change in population size associated with momentum is a linear transformation of the change in mean population age. Momentum and aging are inseparably linked, as both are manifestations of the changes in population size and structure that accompany a transition to long term zero population growth.

## 3.3 MOMENTUM UNDER GRADUAL DECLINES TO STATIONARITY

A good deal of work has been done to explore population momentum in cases where the vital rates decline to replacement level over a number of years. Empirically, Ryder (1975) examined the population of India, calculating a momentum of 1.42 as the initial population of 600 million would grow to 851 million if there were an immediate drop in fertility to replacement. Under the more reasonable assumption that the decline in fertility would span 40 years while life expectancy would rise from 50 to 70 years, however, the ultimate stationary population would be 2.1 billion. Numerically, delays in the achievement of replacement level vital rates, with or without improvements in mortality, can lead to substantial increases in ultimate population size.

Theoretically, Frauenthal (1975) and Mitra (1976) examined different patterns of one-time fertility declines, and Mitra (1987) and Cerone (1996) examined momentum under certain specified patterns of fertility decline. Schoen and Kim (1998) examined the case where the growth rate in the number of births declines linearly to zero. If the initial growth rate of births is $r_B$ and the decline spans Y years, "Keyfitz momentum", i.e. that given by equations (3.1) and (3.2), increases by a factor of $\exp(r_B Y/2)$. In effect, growth at the initial rate continues for half the period of decline.

Li and Tuljapurkar (1999, 2000) derived new mathematical results from renewal theory, summing births from successive generations to find momentum under a variety of age and time patterns of fertility decline. For an initial stable population, and a proportional decline in net maternity over age with a linear decline over time,

$$\Omega_{LT} = \Omega_K[\exp(rY) - 1]/(rY) \tag{3.17}$$

where the subscripts on $\Omega$ refer to the Li-Tuljapurkar and Keyfitz approaches, r is the growth rate of the initial population, and Y is again the length of the period of decline. Equation (3.17) is derived using only first generation births, but appears to be accurate for declines taking up to 40–50 years.

Goldstein (2002) pursued an alternative approach applicable to declines of up to 50 years, finding a momentum, $\Omega_G$, given by

$$\Omega_G = \Omega_K\, R_0^{[Y/50]} \tag{3.18}$$

For an instantaneous decline, $\Omega_G = \Omega_K$, and for a decline phased in over 50 years, $\Omega_G = R_0\ \Omega_K$, the result in Frauenthal (1975). Goldstein and Stecklov (2002) utilized the linear approximation

$$\exp(rY/2) \approx [\exp(rY) - 1]/(rY) \qquad (3.19)$$

to modify $\Omega_{LT}$ and provide the long term expression

$$\Omega_{GS} = \Omega_K \exp(rY/2)\ \{e_L(0)/e_N(0)\} \qquad (3.20)$$

where $e_L(0)$ and $e_N(0)$ are the ultimate stationary and initial population life expectancies and $\Omega_{GS}$ is the Goldstein-Stecklov momentum. Equation (3.20) gave results that corresponded well with conventional long term projections, supporting the argument that such simple analytic projections provide a valuable demographic tool.

Schoen and Jonsson (2003) used a dynamic model that allows monotonic changes in net maternity, is suitable for decline periods of any length, and facilitates analyses of stable-to-stable as well as stable-to-stationary transitions. That model is discussed in detail in Section 7.3 The basic result for gradual declines to stationarity parallels equations (3.17) and (3.19), but adds an additional term that increases exponentially with Y to reflect interactions between changing net maternity and the evolving population age composition.

In short, recent work has shown that, for an initial population growing at rate r, linear declines in net maternity over Y years (say for $Y < 50$) add a factor of about $e^{rY/2}$ to the momentum given by equation (3.1). Differences in the age pattern of net maternity decline do effect momentum, but those effects are generally modest. Differences in the time pattern of fertility declines can be more substantial. Linear declines lead to more growth than exponential declines, as the latter produce greater fertility decreases in the early years.

## 3.4 SPATIAL MOMENTUM

As a change in size associated with a transition to stationarity, momentum is not restricted to birth-death models. In two-sex populations, Schoen (1988a: 191) found a (small) growth effect associated with the difference between the initial and ultimate sex compositions. In multistate models, changes in state composition can lead to population growth, and interregional migration has emerged as the demographic behavior most significantly associated with momentum.

Rogers and Willekens (1978) gave the first analysis of spatial momentum in a multistate context, presenting a multiregional generalization of the relationships in equations (3.1) and (3.2). Extending the Ryder (1975) calculation for India, that analysis saw India's urban fraction, as well as its urban population, increasing substantially during the transition to stationarity.

Several commonly prevailing factors combine to generate spatial momentum, which can be defined as the population growth produced by changes in regional composition during a transition to multistate stationarity. Let us focus on models with the states Urban and Rural. Empirically, interregional migration typically favors the urban areas, as there is more movement to cities than to the countryside. In the present as well as in the past, rural fertility generally exceeds urban fertility. With mortality differentials frequently small, rural areas have a larger rate of natural increase, and that differential in growth is likely to be maintained after the fall in fertility. The ultimate stationary population thus has positive natural increase in the Rural state and negative natural increase in the Urban state, with overall zero growth achieved by net migration from Rural to Urban. As the model population urbanizes during the transition to stationarity, the still overrepresented Rural population continues to generate overall population growth. The transition can be a lengthy process (see subsection 2.4.3), allowing time for a substantial increase in population size.

### 3.4.1 Spatial Momentum in Two Region Models Without Age

Schoen and Kim (1993) analyzed a basic two region (Urban, Rural) model. Let behavior in that model be described by regional rate matrix $\boldsymbol{\mu}_{\mathbf{R}}$, specified by

$$\boldsymbol{\mu}_{\mathbf{R}} = \begin{bmatrix} \rho_U - m_{UN} & m_{NU} \\ m_{UN} & \rho_N - m_{NU} \end{bmatrix} \tag{3.21}$$

where $\rho$ represents a rate of natural increase (birth rate minus death rate), m is a transfer rate, U indicates the Urban state, and N indicates the Rural (or Nonurban) state. The eigenvalues of $\boldsymbol{\mu}_{\mathbf{R}}$ are given by equation 2.33, and here are denoted $r_R$ and $s_R$, with the former the dominant root.

The regional rate matrix can be transformed to a stationary matrix by subtracting $r_R\mathbf{I}$, where $\mathbf{I}$ is the $2 \times 2$ identity matrix. In effect, that subtraction decreases each state's $\rho$ by the overall growth rate implied by $\boldsymbol{\mu}_{\mathbf{R}}$, leaving the transfer rates unchanged and yielding a new rate matrix with zero growth. If the initial urban population is U(0) and the initial rural population is N(0), Schoen and Kim (1993) found that spatial momentum, $\Omega_R$, is given by

$$\Omega_R - 1 = [1/(r_R - s_R)]\{[(\rho_U - r_R)U(0) + (\rho_N - r_R)N(0)]\}/\{(U(0) + N(0)\} \tag{3.22}$$

Equation (3.22) expresses spatial momentum in terms of two factors. The first, $[1/(r_R - s_R)]$, is twice the reciprocal of the force of convergence to stationarity (see equation 2.40). The faster the convergence, the smaller the spatial momentum. The second factor, the ratio of the terms in curly brackets, is the growth rate of the population immediately after the shift to replacement level rates. The faster that growth rate, the greater the spatial momentum. Schoen and Kim (1993) concluded that in a two-region model, spatial momentum leads to population growth when the faster growing state is overrepresented in the initial population.

Schoen (2002) continued that analysis in a work that provides the basis for the rest of this section. To simplify the relationships and eliminate the eigenvalues of $\boldsymbol{\mu}_R$, let $m_{UN}$, the rate of movement from Urban to Rural, be zero. To prevent the rural population from going to zero, it is assumed that $(\rho_N - m_{NU}) > \rho_U$. In this simplified case, spatial momentum, $\Omega_{RR}$, is

$$\Omega_{RR} = \{N(0)/[U(0)+N(0)]\}\{\rho_N - \rho_U\}/\{\rho_N - \rho_U - m_{NU}\} \qquad (3.23)$$

Three elements in equation (3.23) give meaningful behavioral quantities that influence spatial momentum. The first, in the first set of curly brackets, is the initial proportion rural, which varies directly with spatial momentum. The second element, part of both the second and third sets of curly brackets, is the extent to which the rural rate of natural increase exceeds the urban. The *smaller* $\{\rho_N - \rho_U\}$, the greater $\Omega_{RR}$. Heuristically, that is because a smaller difference in rates of natural increase implies a larger ultimate fraction Urban and prolongs the transition to stationarity. The third element is the rural to urban migration rate. The larger $m_{NU}$, the greater spatial momentum. Equation (3.23) thus provides three theoretically based propositions that can be examined in the light of data and in the context of models recognizing both age and region.

### 3.4.2 Decomposing Total Multistate Population Momentum

Let us consider the momentum of a model with 2 regions and $\gamma$ ages. Let the initial, non-stationary PPM of the model be a $2\gamma \times 2\gamma$ matrix, $\mathbf{B}$, with the form

$$\mathbf{B} = \begin{bmatrix} a_{11} & a_{12} & a_{21} & a_{22} & \cdots & a_{\gamma 1} & a_{\gamma 2} \\ b_{11} & b_{12} & b_{21} & b_{22} & \cdots & b_{\gamma 1} & b_{\gamma 2} \\ s_{111} & s_{121} & 0 & 0 & \cdots & 0 & 0 \\ s_{112} & s_{122} & 0 & 0 & \cdots & 0 & 0 \\ 0 & 0 & s_{211} & s_{221} & \cdots & 0 & 0 \\ 0 & 0 & s_{212} & s_{222} & \cdots & 0 & 0 \\ \vdots & \vdots & \vdots & \vdots & \vdots & \vdots & \vdots \end{bmatrix} \qquad (3.24)$$

PPM **B** is a block Leslie matrix (Feeney, 1970), with $a_{ij}$ representing the fertility element (see equation 1.41) for births in state 1 per person in age group i and state j; $b_{ij}$ representing the fertility element for births in state 2 per person in age group i and state j; and $s_{ijk}$ representing the proportion of those initially in age group i and state j who survive to be in state k at age $i+1$. The initial population vector, $\mathbf{x_0}$, has odd numbered elements $x_{j1}$ representing the number of persons in age group j, region 1, and even numbered elements, $x_{i2}$ representing the number of persons in age group i, region 2.

Multistate PPM **B** can be transformed into a matrix with dominant eigenvalue 1 in a number of ways. One is to divide every element of **B** by $\lambda_B$, the dominant root of **B**. However, that procedure changes the survivorship proportions as well as the fertility values. An attractive alternative, which we make use of in the future as well, is to transform the fertility rates to

$$c_{ij}^* = c_{ij}/\lambda_B{}^i \qquad (3.25)$$

where $c_{ij}$ refers to element $a_{ij}$ or $b_{ij}$. Because $\lambda_B$ is raised to the minus ith power in transforming $c_{ij}$ to $c_{ij}^*$, a decline in fertility affects older ages more than younger ages, a realistic pattern of fertility change. That pattern of change was advanced by Sykes (1973), employed in the Coale and Trussell (1974) Model Fertility Schedules, and used in Kim and Schoen (1996).

The initial population can be projected to stationarity by transformed PPM $\mathbf{B}^*$, which is identical to **B** except that it has elements $c_{ij}^*$ instead of $c_{ij}$ in the first two rows. The total momentum of the multi-age/multistate population is then readily found by dividing the total number of persons in that ultimate stationary population by the total number in the initial population. However, that total momentum, $\Omega_T$, reflects the combination of conventional age momentum, $\Omega_A$, spatial momentum, $\Omega_R$, and the interaction between them, $\Omega_I$. To find those individual components, we need to collapse initial PPM **B**.

Let the $\gamma \times \gamma$ Leslie matrix **A** be the age-specific PPM obtained by collapsing **B** over region, using the initial population values to weight the matrix elements involved. With $\lambda_A$ the dominant root of **A**, stationary PPM $\mathbf{A}^*$ can be obtained by applying the transformation used in equation (3.25) to the first row elements of **A**. Age momentum can then be found by projecting the initial age-specific population to stationarity using $\mathbf{A}^*$.

Let the collapsed region-specific PPM be $2 \times 2$ matrix $\mathbf{B_R}$, whose ijth element, $r_{ji}$ is the contribution of each person in state j to the population in state i one interval later. The underlying rate matrix is the $\boldsymbol{\mu}_\mathbf{R}$ shown in equation (3.21), whose dominant root is $r_R$. Matrix $\boldsymbol{\mu}_\mathbf{R}$ can be transformed into stationary rate matrix $\boldsymbol{\mu}_\mathbf{R}{}^*$ by subtracting $r_R\mathbf{I}$. Stationary PPM $\mathbf{B_R}{}^*$ is then obtained from $\boldsymbol{\mu}_\mathbf{R}{}^*$

(see Exercise 7), and $\Omega_R$ is calculated by projecting the initial region-specific population to stationarity using $\mathbf{B_R}^*$. To complete the decomposition, assume that $\Omega_I$ reflects a multiplicative interaction. The value of $\Omega_I$ can then be obtained from

$$\Omega_T = \Omega_A \ \Omega_R \ \Omega_I \tag{3.26}$$

Given the scarcity of actual data regarding the elements of **B**, the three propositions derived in subsection 3.4.1 are examined using plausible but hypothetical data (see Schoen 2002, for details). Fifteen year age intervals are used. Mortality is assumed to be low [$e(0) \approx 78$ years] and fertility moderate, yielding a stable $r \approx .0185$. The initial percents Urban are set at 25% and 37.5%, approximating the urban proportions in the world's less developed areas in 1970 and 1998 (UN Population Division 1999:26). The High urban/rural fertility differential has rural rates 40% above urban; under the Low differential, rural rates are 20% higher. Migration assumptions are based on patterns in Rogers and Castro (1986), and emphasize movements at ages under 30. The rate of rural to urban migration is set at twice the rate of urban to rural. Under the High migration scenario, rural to urban movement, reflected by PPM $\mathbf{B_R}^*$ element $r_{NU}/\lambda_R$, is 0.2 at ages 0–14 and 0.4 at ages 15–29. Under the Medium migration scenario, the values are half that size.

Results for the eight combinations of the initial percent urban, the differential in urban/rural fertility, and the urban/rural migration level are shown in Table 3.1. Total momentum varies from 1.376 to 1.656. Most of the growth is attributable to age momentum, and interaction factors are rather small. Spatial momentum ranges from 0.950 to 1.226. In 6 of the 8 cases shown, spatial momentum leads to growth of at least 8% beyond that due to age momentum. All three theoretical propositions are confirmed. In comparisons between values where only one factor changes, spatial momentum is always higher when the initial proportion rural is higher, when the urban/rural fertility differential is lower, and when the pace of rural to urban migration is higher. Moreover, spatial momentum is quite sensitive to the difference between the initial and ultimate fractions urban, as $\Omega_R$ is usually close to 1 plus the increase in the proportion urban.

An analysis of spatial momentum using actual (though partially estimated) data is possible in the case of Mexico 1970 [see Schoen 2002 for details]. The demographic transition began in Mexico around 1970, and the country has experienced considerable internal and international migration. Calculations using the female population were made following the procedures outlined above. The results show an increase in the proportion urban from .596 to .820, and indicate the following patterns of growth over the transition to stationarity:

| Momentum Type | Value |
|---|---|
| Total | 2.031 |
| Age | 1.721 |
| Spatial | 1.148 |
| Interaction | 1.028 |

Total momentum more than doubles the initial population size, while spatial momentum alone accounts for 15% growth.

Although it has received little attention, spatial momentum is a significant demographic phenomenon. Many contemporary countries have a relatively low percent urban, a modest urban/rural fertility differential, and considerable rural-to-urban migration. As those are the conditions that generate spatial momentum, interregional redistribution can have a substantial impact on future world population growth.

*Table 3.1. Momentum and Ultimate Proportion Urban Values for Hypothetical Populations By Initial Percent Urban, Urban/Rural Fertility Differential, and Level of Urban/Rural Migration*

| | Momentum | | | | Ultimate Proportion Urban |
|---|---|---|---|---|---|
| Characteristics | Total | Age | Spatial | Interaction | |
| A. High urban/rural fertility differential | | | | | |
| 1. 25% initially urban | | | | | |
| a. high migration | 1.656 | 1.393 | 1.207 | .985 | .468 |
| b. medium migration | 1.572 | 1.393 | 1.094 | 1.032 | .337 |
| 2. 37.5% initially urban | | | | | |
| a. high migration | 1.478 | 1.345 | 1.084 | 1.013 | .462 |
| b. medium migration | 1.376 | 1.345 | .950 | 1.076 | .329 |
| B. Low urban/rural fertility differential | | | | | |
| 1. 25% initially urban | | | | | |
| a. high migration | 1.639 | 1.383 | 1.226 | .967 | .519 |
| b. medium migration | 1.650 | 1.383 | 1.149 | 1.038 | .392 |
| 2. 37.5% initially urban | | | | | |
| a. high migration | 1.505 | 1.341 | 1.121 | 1.001 | .520 |
| b. medium migration | 1.493 | 1.341 | 1.019 | 1.092 | .394 |

*Source:* Adapted from Schoen (2002, Table 1).

## 3.5 SUMMARY

Population momentum generally refers to the growth that follows a sudden drop in vital rates to replacement level, and reflects the reproductive potential of the initial population. It is a demographic phenomenon that is likely to be responsible for most of the future growth of the world's population. Momentum can be defined as the ratio of the total size of the ultimate stationary population to the total size of the initial population. For contemporary populations, momentum ($\Omega$) can easily be 1.5 to 1.7. In birth-death models, momentum can be calculated as the product of 3 factors: the initial birth rate, the ultimate life expectancy, and the stable equivalent number of births. Essentially all of the growth in birth-death models that is attributable to momentum comes at ages over 30, inextricably linking momentum and population aging. Theoretically and empirically, momentum is a simple linear function of the change in mean population age over the transition to stationarity. If the fall to replacement level fertility is not immediate but takes place over Y years, momentum increases by a factor that approximates continued growth at the initial rate for Y/2 years.

Momentum also arises in multistate models, where spatial momentum reflects the additional population growth attributable to changes in regional composition. In two region models, spatial momentum can be sizeable when the faster growing region is overrepresented in the initial population. Since that characterizes the urban and rural areas of much of the contemporary developing world, spatial momentum is likely to play a significant role in global population growth over the next century.

## 3.6 EXERCISES

[See Appendix B for selected answers]

1. A power series often seen in mathematical demography is

$$e^x = 1 + x + (1/2!)\,x^2 + (1/3!)\,x^3 + \ldots$$

Using that series to linear terms and the (rough) approximation that mean age at childbearing $\mu$ equals Lotka's $\mathrm{T}$, verify equation (3.3) by showing that Q in equation (3.2) is about $(R_0)^{-1/2}$.

2. Assume that a stable population growing at intrinsic rate r experiences a sudden fall in fertility to replacement level. Show that
   a. the age at which the initial stable and ultimate stationary populations have the same proportion, $y_p$, is given by

$$y_p = \ln[be(0)]/r$$

b. the age at which the initial stable and ultimate stationary populations have the same number of persons, $y_N$, is given by

$$y_N = \ln[be(0)/Q]/r$$

3. Consider a 3 age group Leslie PPM with a double root. Show that its momentum is given by

$$\Omega = b\ e(0)\ \{(1+[c']^{1/2}/\zeta)/(1+[c']^{1/2})\}^2$$

where c' is defined following equation (2.22) and $\zeta$ is the growth parameter of the initial stable population.
4. Consider the $3 \times 3$ PPM (with 15 year age groups)

$$\mathbf{AA} = \begin{bmatrix} .36 & .864 & .1728 \\ 1 & 0 & 0 \\ 0 & 1 & 0 \end{bmatrix}$$

Verify that the dominant root of **AA** is 1.2. Assume initial population $\mathbf{x}' = (1, .83333, .69444)$.

a. Assume a proportional fall in fertility to replacement level. Verify that

$$\mathbf{v}' = (.535911, .397791, .066298) \text{ and } Q = .913444$$

b. Assume that fertility falls to replacement via the Sykes transform of equation (3.25).
Verify that the PPM first row elements are (.3, .6, .1); $\mathbf{v}' = (.555556, .388889, .055556)$; and $Q = .918210$.
c. Compare the initial population under exact age 30 with that in the stationary populations implied by the 2 patterns of fertility fall. Evaluate equation (2.27) in this context.

5. Assume that initial population $\mathbf{x}' = (0, 0, 1)$ is projected into the future by a series of stationary level PPMs. Those PPMs need not be the same, but all must have a dominant root of 1. What is the largest possible size of the projected population?
6. The matrix **B** in equation (3.24) could also be transformed into a PPM with dominant root 1 by subtracting $\lambda_B\mathbf{I}$, with **I** the $2\gamma \times 2\gamma$ identity matrix. Is that a viable approach?
7. Rate matrix $\boldsymbol{\mu}_\mathbf{R}$ in equation (3.21) is transformed to a rate matrix with dominant eigenvalue 0 by subtracting $r_R\mathbf{I}$ (with **I** the $2 \times 2$ identity matrix). Show that $\mathbf{B_R}$, the PPM that follows from $\boldsymbol{\mu}_\mathbf{R}$, is transformed to PPM $\mathbf{B_R}^*$ with dominant root 1 by dividing each element of $\mathbf{B_R}$ by $\exp(r_R)$.

# CHAPTER 4

# DEMOGRAPHIC CHANGE AT THE MARGIN

## 4.1 INTRODUCTION

Life tables and stable populations depict the implications of specified, fixed rates for population size and composition. To relax the fixed rate constraint, we begin by examining a change in rates in the context of fixed rate models. We then consider age-dependent growth, and explore connections between different populations with the same vital rates. The objective is to understand how demographic rates influence population structures. Its pursuit leads to a fundamental principle underlying population dynamics, derivatives of demographic measures, and a spectral decomposition of population growth.

## 4.2 MARGINAL CHANGE IN FIXED RATE MODELS

In the basic life table, equation (1.5) shows how the number of survivors to any age depends on the number of persons at the initial age and the force of mortality function between those ages. Now let us consider the changes in survivorship that result from some patterned changes in mortality. While not particularly realistic, the simplest form of change to analyze is an additive change in the force of mortality at all ages, i.e.

$$\mu^*(x) = \mu(x) + c \qquad (4.1)$$

where the asterisk indicates the new force of mortality and c is the constant change at all ages.

The new survivorship function, $\ell^*(x)$, is then

$$\ell^*(x) = \ell(x)\exp[-cx] \qquad (4.2)$$

indicating that the difference in survivorship widens exponentially over age.

A more realistic form of change is when the risk of death changes proportionally at all ages, or

$$\mu^*(x) = \mu(x)[1+k] \tag{4.3}$$

Under that form of mortality change

$$\ell^*(x)/\ell(x) = [\ell(x)/\ell(0)]^k \tag{4.4}$$

Keyfitz (1985) showed that under equation (4.3) life expectancy at birth changes by

$$\Delta\ e(0) = -k\ e(0)\ H \tag{4.5}$$

where H is the entropy of the survivorship distribution, defined as

$$H = -\int p(x)\ \ln\ [p(x)]\ dx/e(0) \tag{4.6}$$

Since entropy is well below 1 in low mortality populations, the entropy factor reduces the impact of proportional mortality changes on longevity. For example, consider a 10% decline in death rates at all ages. If $H = 0.2$, a reasonable value, and the initial e(0) is 70 years, life expectancy increases by only $(0.1)(70)(0.2) = 1.4$ years.

### 4.2.1 Changes in Stable Population Measures

In the context of stable populations, increases in mortality can make the population older or younger, depending on the pattern of mortality and the nature of the change. The additive change in equation (4.1) has been termed a "neutral" change in mortality as it leaves age composition unchanged (see Exercise 1). Deaths can occur at any age, but fertility directly impacts the lower end of the age distribution as births always occur at age zero. It follows that fertility declines lower intrinsic growth rate r, lower the fraction at the younger ages, and raise the fraction at the older ages.

To analyze how the age composition of a stable population changes with r, we can use equation (1.35) and examine

$$d\ \ln\ c(x)/dr = [1/c(x)]\ (d/dr)\ c(x) = (d/dr)\ \ln[b\ e^{-rx}\ p(x)] \tag{4.7}$$

The probability of survival does not vary with r, and equations (1.32) and (1.33) imply that [(d/dr) ln b] is the mean age of the stable population ($Y_r$). Hence equation (4.7) can be rewritten

$$(d/dr)\ c(x) = c(x)\ [Y_r - x] \tag{4.8}$$

an expression both simple and meaningful. Age composition pivots around $Y_r$. A decrease in r decreases c(x), in proportion to its initial size, at all ages below the mean population age. Since the change is greater the further an age is from $Y_r$, the greatest impacts are at age zero and the highest ages. For example, if r decreases by 0.01 in a population with a mean age of 30, c(0) would change by c(0)[−0.01][30–0] or decrease by [0.3c(0)]. Thus a one percentage point drop in r lowers the proportion at age 0 by 30%.

Arthur (1984) used functional calculus to explore linkages between demographic measures under any pattern of change in vital rates. Let $\delta y$ be the change in function y brought about by perturbation $\delta x$ in function x. Then the linear approximation of the effect on survival of any set of changes in the force of mortality, $\delta\mu$, is given by

$$\delta p(x) = -p(x) \int_0^x \delta\mu(a)da \qquad (4.9)$$

If the change in $\mu$ is proportional at all ages, equation (4.9) provides a linear approximation to the result in equation (4.4).

Arthur (1984) showed how the power of functional differentiation can illuminate the nature of changes in stable population measures. Changes in the intrinsic growth rate are related to changes in fertility and survivorship by

$$\delta r = \{\int e^{-rx}[\delta p(x)f(x) + p(x)\delta f(x)]\, dx\}/A^* \qquad (4.10)$$

The impact on growth is thus scaled by the stable population mean age at childbearing, so that equivalent changes in fertility or survivorship have less impact when $A^*$ is larger.

The complex effects of mortality change on age composition are the result of three different effects, as

$$\delta c(x)/c(x) = \delta p(x)/p(x) - b\int e^{-ra}\, \delta p(a)da + [(Y_r - x)/A^*]\int e^{-ra}\, f(a)\, \delta p(a)da \qquad (4.11)$$

The second term on the right side of equation (4.11) tends to be approximately constant, while the first and third terms tend to offset one another. In the West up to around 1970, 20th century mortality declines at ages under 5 were proportionately greater than at most other ages. The third term in equation (4.11) thus had a greater effect than the first, and mortality declines increased the proportion at younger ages, making the population younger. In recent decades, as child mortality has reached low levels and there have been sustained mortality improvements at older ages, the first term has dominated the third and mortality change has contributed to population aging.

The effects of fertility on age composition are simpler. Arthur (1984) found

$$\delta c(x)/c(x) = [(Y_r - x)/A^*] \int e^{-ra}\, p(a)\, \delta f(a) da \qquad (4.12)$$

As the effect of an increase in fertility is to increase r, equation (4.12) is somewhat similar to equation (4.8), but it is now clear that the impact of fertility is scaled by the stable population mean age at childbearing.

### 4.2.2 Changes in Stable Population Characteristics

Coale (1972, Ch 2) examined stable populations where the proportion of persons with a given characteristic, or who experience a given event, varies with age. Let g(x) be the rate (or proportion) at age x, and let G be the overall rate (or proportion of persons with that characteristic) in the stable population. Then

$$G = \int g(x)\, c(x)\, dx \qquad (4.13)$$

To see how G varies with r, differentiate using equation (4.8), which yields

$$dG/dr = G[Y_r - Y_g] \qquad (4.14)$$

where $Y_g$ is the mean age of the population with characteristic g. At the margin, the proportional change in G is a simple difference between the mean age of the entire stable population and that of the fraction with characteristic g. Equation (4.8) can be seen as the special case of equation (4.14) where the specified characteristic is being age x. Coale (1972) noted that an extremum of G occurs when $Y_r = Y_g$. That extremum is a maximum when the variance of the population with characteristic g (denoted by $\sigma^2{}_g$) is less than the variance of the stable population age distribution ($\sigma^2{}_Y$), and a minimum when $\sigma^2{}_g > \sigma^2{}_Y$. Using equation (4.14), Coale (1972) verified an earlier result of Lotka, showing that a stable population has its minimum crude death rate when $Y_r = 1/b$. Equation (4.14) applies to multistate stable populations as well.

Preston (1982) extended that line of analysis to examine how changes in growth differentially affect period measures in stable populations and the experience of cohorts in those stable populations. Changes in fertility affect period measures, but not the life course of any cohort. Neutral changes in mortality [i.e. those of equation (4.1) and, to a fair extent, those similar to equation (4.3)] affect individual life cycles but do not appreciably impact the composition of any period. Let $G_P$ be the proportion of persons in the stable population with characteristic g, and $G_L$ be the analogous cohort (stationary population) proportion. A change in r changes $G_P$ by

$$\Delta G_P/\Delta r = G_P(Y_L - Y_{GL}) \quad (4.15)$$

where $Y_L$ is the mean age of the stationary population and $Y_{GL}$ is the mean age of those with attribute g in the stationary population. At the margin, a neutral change in mortality of size c has the same impact on cohort measures, i.e.

$$\Delta G_L/\Delta c = G_L\ (Y_L - Y_{GL}) \quad (4.16)$$

Preston (1982) used those relationships to examine situations where population characteristics (e.g. occupational structures or consumption levels) are constrained and force alterations in life cycles. For example, based on United States data for 1977, a 0.01 increase in r would decrease the stable population proportion living in nursing homes from 0.0110 to 0.0065, a 41% decline. The mean age of those in nursing homes was 80.2 years, while the mean age of the stationary population was 38.9, 41.3 years younger. Those changes are consistent with equation (4.15), multiplied by $\Delta r/G_P$, as [.0065–.0110]/(.0110)=(.01)[38.9–80.2] = −.41. The same proportional decline, but in the cohort fraction of life lived in nursing homes, would follow from a 0.01 increase in the force of mortality at all ages.

## 4.3 RECOGNIZING AGE-DEPENDENT GROWTH

In a stationary population, equation (1.3) indicates that the proportional change in the number of survivors at age x is $-\mu(x)$. In a stable population with $n(x) = e^{-rx}p(x)$ persons at age x, the proportional change in n(x) over age is

$$d[\ln n(x)]/dx = -\mu(x) - r \quad (4.17)$$

Movement down the age column in a stable population results in fewer persons for two reasons: mortality and growth in cohort size.

Preston and Coale (1982), building on the analysis of Bennett and Horiuchi (1981), generalized that relationship to any population. Let N(x,t) be the number of persons age x at time t, and let r(x,t) be defined as

$$r(x,t) = [1/N(x,t)]\ \partial N(x,t)/\partial t \quad (4.18)$$

Age-dependent growth rate r(x,t) is thus the proportional change in population size over *time*. It follows that

$$[1/N(x,t)]\ \partial N(x,t)/\partial x = -\mu(x,t) - r(x,t) \quad (4.19)$$

where the force of mortality can also vary with time. The population in year t can then be found by integrating equation (4.19), yielding

$$N(x,t) = N(0,t)\ \exp[-\int_0^x r(a,t)da]\ p(x,t) \tag{4.20}$$

where N(0,t) = B(t) is the number of births occurring during year t.

Equation (4.20) is a significant extension of stable population formalism to any study population. Dividing both sides of the equation by N(t), the total size of the year t population, yields a relationship analogous to equation (1.35) for the age composition of a stable population. The core of the Preston-Coale "variable-r" approach is the ability to replace $e^{-rx}$ by $\exp[-\int r(a,t)da]$, where the integration goes from 0 to x. The sum over the age-specific r(a,t) reflects past differences in birth cohort size and age-specific mortality, allowing p(x,t) to reflect the death rates of year t alone. Preston and Coale (1982) found variable-r expressions that extended all of the basic stable population relationships. Horiuchi and Preston (1988: p429,438) saw the r(x,t) as "the legacy of past population dynamics", though they acknowledged that "[a]ge-specific growth rate profiles cannot be unambiguously interpreted."

Those same general relationships were then derived by Arthur and Vaupel (1984) in terms of an age-time Lexis surface. That analysis defined the proportional change over age, $\nu(x,t)$, by

$$\nu(x,t) = -[1/N(x,t)]\ \partial N(x,t)/\partial x \tag{4.21}$$

The fundamental local identity linking change over age and time could then be written

$$\mu(x,t) = \nu(x,t) - r(x,t) \tag{4.22}$$

a form equivalent to the classical differential equation of von Förster (1959).

In discrete applications, the value of the age-dependent growth function over the interval from t to t + s can be found from

$$r(x, n, t+s) = (1/s)\ \ln\ [N(x, n, t+s)/N(x, n, t)] \tag{4.23}$$

where r(x, n, t + s) is the growth rate of the (x, n) age interval over the time interval ending at time t + s, and N(x, n, t) is the population aged x to x + n at time t.

Preston and Coale (1982) described how the variable-r approach could be used to estimate a wide range of demographic measures. Kim (1986) further developed the age-dependent growth relationships in discrete form, but questioned the approach's usefulness in demographic estimation because of the data needed to calculate the r(x, t). Despite that significant limitation, the variable-r approach has contributed to an integrated system for demographic estimation from two age distributions (Preston 1983), and has proven to be useful in situations where age intervals and time intervals are unequal.

## 4.4 CONNECTIONS BETWEEN POPULATIONS HAVING THE SAME VITAL RATES

The same vital rates can occur in populations with different age or state compositions. The most commonly encountered examples are observed, stable, and stationary populations based on the same vital rates. To explore the connections between demographic measures in such populations, we examine points of equality that have been termed "crossover" points.

A crossover point is a point where demographic density distributions from two populations with the same rates intersect (Kim and Schoen 1993). Consider birth density distributions, i.e. the proportional contributions to the total number of births from persons at each age, arising from a stationary and a stable population with the same fertility and mortality. If those distributions are not identical, they must intersect because the area underneath each curve is one (by definition). At such a crossover point, say $B_1$, the birth densities are equal, so

$$p(B_1)f(B_1)/\int p(x)f(x)dx = \exp(-r\ B_1)\ p(B_1)f(B_1)/\int e^{-rx}\ p(x)f(x)dx \qquad (4.24)$$

Cancelling, using equations (1.29) and (1.30), and rewriting gives

$$R_0 = \exp(r\ B_1) \qquad (4.25)$$

Equation (1.38) indicates that crossover point $B_1$ is Lotka's Ŧ.

More generally, stationary and stable population crossovers of distributions based on characteristic Z lead to the relationship

$$\text{Total stationary cases of } Z = (\text{Total stable cases of } Z)\ \exp(r\ X_Z) \qquad (4.26)$$

where $X_Z$ is the point where the stationary and stable density distributions of Z cross over. For example, in equation (4.25), the unit birth in the stable population, adjusted by stable growth over Ŧ years, yields the $R_0$ births in the cohort (or stationary population). Characteristic Z can reflect any of a number of demographic distributions, including those of deaths, reproductive values, population numbers, marriages, and divorces.

Preston (1975; 1978) related stationary to stable population summary measures for a broad range of demographic events, e.g. how the probability a person ever marries relates to the stable population period ratio of first marriages to births. That relationship can be written

$$\mathrm{Prob}(M_1) = [M_1/B]\ \exp(r\ A_{M1}) \tag{4.27}$$

where $\mathrm{Prob}(M_1)$ is the cohort probability of first marriage, $M_1$ is the number of period first marriages in a stable population with the same rates, B is the number of births in the stable population that year, and $A_{M1}$ is the stable population average age at first marriage.

As Preston (1978) noted, that approach relates to work by Keyfitz (1977: p126), which showed that Lotka's ∓ could be expressed in terms of an infinite series, specifically

$$\mp = \mu - r\ \sigma^2/2 + r^2\ \mu_3/6 - \ldots \tag{4.28}$$

where $\mu$, $\sigma^2$, and $\mu_3$ are the first, second and third cumulants of the cohort birth distribution. [The first two cumulants of a distribution are its mean and variance; the third cumulant is the third moment around the mean, and is related to skewness (cf. Kendall and Stuart 1958: Vol. I, p67–74).] The relationships in Preston (1975; 1978) are essentially linear approximations. Using additional terms from the series in Keyfitz (1977) leads to the exact, crossover, value. In equation (4.27), that would be obtained by replacing $A_{M1}$ by the age at which the stationary and stable first marriage density distributions intersect, i.e. by $\mu_N - r\sigma^2{}_N/2 + r^2\mu_{3N}/6 - \ldots$ , where $\mu_N, \sigma^2{}_N, \mu_{3N}$, and so on, are the first, second, third, etc. cumulants of the stationary population distribution of first marriages.

Now let us consider crossover points involving any observed population. Using variable-r notation, the stable-any birth crossover point, $B_3$, can be written

$$\exp[-\int_0^{B_3} r(a)da]p(B_3)f(B_3)/\int \exp[-\int_0^{x} r(a)da]\ p(x)f(x)dx$$
$$= \exp(-rB_3)\ p(B_3)f(B_3)/\int e^{-rx}\ p(x)f(x)dx \tag{4.29}$$

Simplifying as before, noting that the variable-r characteristic equation also equals 1 (see Exercise 4), taking logs, and rewriting yields the result in Kim and Schoen (1993), i.e.

$$r = (1/B_3) \int_0^{B_3} r(a)da \tag{4.30}$$

The average growth rate up to age $B_3$ is Lotka's r. Put differently, the previous $B_3$ years of growth in the stable population defined by the prevailing rates is the same as the previous $B_3$ years of age-dependent growth in the observed population. It follows that if $r = 0$, age $B_3$ is the age up to which there is no growth over the transition to stationarity, and thus $B_3$ is the correct value for G in equation (3.7). With an actual population, however, there may be multiple crossover values, and those values must be found by graphical, approximate, or "brute force" methods (cf. Kim and Schoen 1993). No series analogous to equation (4.28) is known. In principle if not in convenient practice, crossover-age based adjustments can be found to transform a summary measure in one population to the corresponding summary measure in any other population with the same vital rates.

## 4.5 MARGINAL CHANGE IN DYNAMIC POPULATIONS

### 4.5.1 Derivatives of Aggregate Demographic Measures

Schoen and Kim (1992) showed how the age-dependent growth function can be used to determine how a variety of measures change over time. Rewriting equation (4.18) gives

$$\partial N(x,t)/\partial t = r(x,t)\, N(x,t) \tag{4.31}$$

or that at time t the number of persons at age x grows at rate r(x, t). Growth at age zero can also be expressed using the renewal equation. By definition, the number of births at time t is

$$B(t) = N(0,t) = \int N(x,t)\, f(x,t)\, dx \tag{4.32}$$

Differentiating with respect to time using equation (4.31) and dividing by B(t) yields

$$\begin{aligned} r(0,t) &= R_B(t) = \partial \ln B(x,t)/\partial t \\ &= \{\int r(x,t)\, N(x,t)\, f(x,t)\, dx + \int N(x,t)\, [\partial f(x,t)/\partial t]\, dx\}/B(t) \end{aligned} \tag{4.33}$$

with $R_B(t)$ denoting the rate of increase in births at time t. If fertility is constant over time, equation (4.33) simplifies to

$$R_B(t) = \int r(x,t)\, N(x,t)\, f(x)\, dx/B(t) \tag{4.34}$$

or the average of the growth rates weighted by the age distribution of births.

Equation (4.31) leads to the time derivatives of a number of demographic functions. With N(t) the total number alive at time t,

$$R_N(t) = \partial \ \ln \ N(t)/\partial t = \int N(x,t) \ r(x,t) \ dx/N(t)$$
$$= \int r(x,t) \ c_N(x,t) \ dx \qquad (4.35)$$

where $R_N(t)$ is the crude rate of natural increase at time t and $c_N(x,t)$ is the time t population proportion at age x. Equation (4.35) directly parallels equation (4.34), and shows that the overall growth rate is the composition weighted average of the age-specific growth rates. The population age composition itself changes according to

$$\partial c_N(x,t)/\partial t = (\partial/\partial t) \ [N(x,t)/N(t)] = c_N(x,t) \ [r(x,t) - R_N(t)] \qquad (4.36)$$

The proportional change in the population fraction at age x is thus the difference between growth at age x and overall growth.

Many demographic summary measures are age-aggregated functions of the form

$$G(t) = \int Z(x,t) \ g(x) \ dx / \int Z(x,t) \ dx \qquad (4.37)$$

where g(x) can be any function of age and Z(x, t) is any variable that satisfies

$$\partial Z(x,t)/\partial t = r(x,t) \ Z(x,t) \qquad (4.38)$$

Accordingly, Z(x, t) can refer to persons, births, deaths, or other demographic distributions. For example, if $Z(x,t) = N(x,t)$ and $g(x) = \mu(x)$, then G(t) is a crude death rate. If $Z(x,t) = N(x,t)f(x)$ and $g(x) = x$, then G(t) is an average age at childbearing.

To examine change in G(t), we differentiate equation (4.37) and find

$$\partial G(t)/\partial t = \int r(x,t) \ Z(x,t) \ g(x) \ dx / \int Z(x,t) \ dx$$
$$- G(t) \ [\int r(x,t) \ Z(x,t) \ dx / \int Z(x,t) \ dx] \qquad (4.39)$$

The second ratio of integrals on the right hand side of equation (4.39) is the overall rate of growth in Z at time t, and can be denoted $R_Z(t)$. The expression

$$\int r(x,t) \ Z(x,t)g(x) \ dx / \int Z(x,t)g(x) \ dx$$

is the overall growth rate of Zg, and can be denoted $R_{Zg}(t)$. Hence equation (4.39) can be rewritten

$$\partial G(t)/\partial t = G(t) \ [R_{Zg}(t) - R_Z(t)] \qquad (4.40)$$

Equation (4.40) shows that the proportional change in G(t) is the difference between the growth rate of Zg and the growth rate of Z. Although it is similar to equation (4.14), the Coale (1972) change relationship, equation (4.40) is more general. Equation (4.14) gives the marginal change in the composition of a stable population when it begins to move toward stability based on a different intrinsic growth rate. Equation (4.40) gives the marginal change over time in G(t) when any given initial population experiences a change in demographic rates.

Extending an insight in Preston, Himes and Eggers (1989), Schoen and Kim (1992) observed that equation (4.39) can be written in terms of covariances. If E(X) is the expected value of random variable X, then the covariance of random variables X and Y can be written (Hanushek and Jackson 1977: p331)

$$\mathrm{cov}(X, Y) = E(XY) - E(X)\,E(Y) \qquad (4.41)$$

Let $\mathrm{cov}_Z(r, g)$ be the covariance between r(x, t) and g(x), where those variables are weighted by Z(x, t). Then equation (4.41) implies

$$\partial G(t)/\partial t = \mathrm{cov}_Z(r, g) \qquad (4.42)$$

The marginal change in aggregate measure G(t) is thus the Z-weighted covariance between the age graded variable g(x) and age-specific growth. The more that large values of g(x) are associated with large values of r(x, t), the more G(t) increases. That result is not surprising, but it is significant because it specifies and quantifies the mechanism underlying changes in aggregate measures.

To take a specific example, let Z(x, t) = N(x, t) and g(x) = f(x). Then G(t) is the crude birth rate and

$$\partial BR(t)/\partial t = BR(t)\,[R_B(t) - R_N(t)] = \mathrm{cov}_N(r, f) \qquad (4.43)$$

The proportional change in the crude birth rate is the difference between the rate of increase in births and the overall rate of natural increase. Alternatively, that proportional change is the population weighted covariance between the rates of age-specific growth and fertility.

Preston, Himes and Eggers (1989) derived an expression for the change in the mean age of a population (see Exercise 6). An alternative, but algebraically equivalent, expression follows from setting Z(x, t) = N(x, t) and g(x) = x, i.e.

$$\partial Y_N(t)/\partial t = Y_N(t)\,[R_{Nx}(t) - R_N(t)] = \mathrm{cov}_N(r, x) \qquad (4.44)$$

where $Y_N(t)$ is the average age of the observed population at time t and $R_{Nx}(t)$ is the overall rate of increase in Nx (i.e. population times age). Equation (4.44) shows that a population ages to the extent that age-specific growth covaries with age. For reference, Table 4.1 gives time derivatives for a number of common demographic functions.

*Table 4.1. Time Derivatives of Population Functions*

| Function | Derivative expressed in terms of: Covariance | Growth rates |
|---|---|---|
| 1. Observed population (N) | — | $r(x)\,N(x)$ |
| 2. Age composition ($c_N$) | $cov_N(\delta^*(a-x), r(x))$ | $c_N(a)[r(a)-R_N]$ |
| 3. Birth rate (BR) | $cov_N(f(x), r(x))$ | $BR[R_B-R_N]$ |
| 4. Death rate (DR) | $cov_N(\mu(x), r(x))$ | $DR[R_D-R_N]$ |
| 5. Rate of population increase ($R_N$) | $cov_N(\{f(x)-\mu(x)\}, r(x))$ | $BR\ R_B - DR\ R_D - R_N{}^2$ |
| 6. Ratio BR/DR | $cov_N(\{f(x)-\mu(x)BR/DR\},\ r(x))/DR$ | $BR(R_B-R_D)/DR$ |
| 7. Mean population age ($Y_N$) | $cov_N(x, r(x))$ | $Y_N[R_{Nx}-R_N]$ |
| 8. Mean age at childbearing ($A_N$) | $cov_B(x, r(x))$ | $A_N[R_{Bx}-R_B]$ |
| 9. Mean age at death ($Y_\delta$) | $cov_D(x, r(x))$ | $Y_\delta[R_{Dx}-R_D]$ |

*Notes*: Time index (t) is not shown. The subscript of cov refers to the number of persons (N), births (B), or deaths (D) ; the function $\delta^*(a-x)$ is 1 when $a = x$ and 0 otherwise; $R_{Nx}$, $R_{Bx}$, and $R_{Dx}$ denote the overall rate of increase in persons times age, births times age, and deaths times age, respectively.
*Source:* Adapted from Schoen (1992). Reprinted with permission from *Population Index*, Office of Population Research, Princeton University, (c) 1992.

### 4.5.2 Marginal Change and Stabilization

Convergence to stability is the result of a series of marginal changes that occur as a population is exposed to a fixed set of rates. To look at a given step in the process, we follow Schoen and Kim (1991), and start by letting Q(t) be the stable equivalent number of births at time t [see equation (2.7)]. If the rates are constant for even a momentary period of time, Q(t) grows exponentially at the stable r implied by those rates, hence

$$dQ(t)/dt = rQ(t) = \int r(x,t)N(x,t)v(x)dx \qquad (4.45)$$

where v(x) are the reproductive contributions based on the prevailing rates [see equation (2.8)]. With q(x, t) the proportional contribution of the time t population age x to Q(t) [see equation (2.9)], we find

$$dq(x,t)/dt = q(x,t)\ [r(x,t) - \textstyle\int r(x,t)q(x,t)dx] \tag{4.46}$$

Using equations (4.45) and (2.7), the integral in equation (4.46) is

$$r = \textstyle\int r(x,t)q(x,t)dx \tag{4.47}$$

a striking relationship that equates stable growth under the prevailing rates to the average of the age-specific growth rates weighted by q(x, t). Equation (4.46) then becomes

$$dq(x,t)/dt = q(x,t)\ [r(x,t) - r] \tag{4.48}$$

showing that at any time t the proportional contribution to the current stable equivalent from age x changes in proportion to the difference between age-specific growth at age x and the stable growth implied by prevailing rates.

At time t, the distance between the observed population and the stable equivalent population implied by the prevailing rates is given by the Kullback distance, K(t), of equation (2.13). Defining the momentum of persons age x [or between ages x and x + dx] at time t by $\Omega(x, t)$, we can write

$$\Omega(x,t) = S(x,t)/N(x,t) \tag{4.49}$$

and express the Kullback distance as

$$K(t) = -\textstyle\int q(x,t)\ \ln\ \Omega(x,t)\ dx \tag{4.50}$$

At the margin, the change in the Kullback distance is

$$dK(t)/dt = K(t)\ [r_K(t) - r] \tag{4.51}$$

where

$$r_K(t) = -\textstyle\int r(x,t)\ q(x,t)\ \ln\ \Omega(x,t)\ dx/K(t) \tag{4.52}$$

or $r_K(t)$ is the average of age-specific growth weighted by each age's proportional contribution to the Kullback distance. Applying equation (4.42), we have the result in Schoen and Kim (1991)

$$dK(t)/dt = -\mathrm{cov}_q(r, \ln \Omega) \tag{4.53}$$

The decline in the Kullback distance at any time reflects the q-weighted covariance between age-specific growth and the natural logarithm of age-specific momentum. Convergence proceeds more rapidly when ages with larger growth rates are small relative to their stable equivalent.

The result in equation (4.53) is broadly applicable because the prevailing rates are assumed fixed for only an instant. In the interpretation of Schoen and Kim (1991: p460) a fundamental principle of population dynamics is that every population always moves toward the stable equivalent implied by the prevailing vital rates, and does so by an amount equal to the q-weighted covariance between log momentum and age-specific growth. As the rates change the target stable population changes, but the underlying dynamic principle remains the same.

## 4.6 A SPECTRAL DECOMPOSITION OF MARGINAL CHANGE

The discrete model can provide a different perspective on the relationships between r(x, t), $\Omega(x,t)$, and q(x, t), three functions central to population change at the margin. Using the definition in Kim (1986), discrete age-specific growth is given by

$$\exp[r_{jt}] = N_{jt}/N_{j,t-1} \tag{4.54}$$

where $r_{jt}$ is the growth rate of the jth age group between time points t–1 and t, and $N_{jt}$ is the number of persons in the jth age group at time t. The unit of time is the number of years in the interval between t–1 and t, which is also the number of years in each age group. Using that definition, Schoen and Kim (1992) derived a discrete form of equation (4.39), and provided first differences for the change in a number of population functions (see Exercise 9).

Given a study population and a set of age-specific vital rates, it is possible to spectrally decompose that population, i.e. divide it into components related to the roots of the associated population projection (Leslie) matrix [cf. Keyfitz, 1977: Ch 3]. The number of persons in the ith spectral component may be negative or complex, but that component grows over time according to the ith root of the Leslie matrix. In the spectral decomposition, let $N_{ijt}$ be the number in the ith component and jth age group at time t. Component 1 is the component associated with the dominant root, and $N_{jt} = \Sigma_i N_{ijt}$. Following the development in Schoen and Kim (1992), we have

$$N_{ijt} = \lambda_i^t Q_{i0} u_{ij} \tag{4.55}$$

where $\lambda_i$ is the ith eigenvalue of the Leslie matrix, $Q_{it}$ is the stable equivalent number of births in the ith component at time t, and $u_{ij} = \lambda_i^{-(j-1)} p_j$ is the jth element of the right eigenvector associated with $\lambda_i$. Survivorship fraction p is defined in equation (1.43), and $Q_{i0}$ can be written

$$Q_{i0} = \Sigma_j \ N_{j0} \ v_{ij} \tag{4.56}$$

where $v_{ij}$ is the jth element of the left eigenvector associated with $\lambda_i$. In terms of the $b_j$ fertility elements of equation (1.41),

$$v_{ij} = \Sigma_{k=j} \ \lambda_i^{-k} p_k \ b_k/u_{ij} \ \kappa_i \tag{4.57}$$

where $\kappa_i$ is the stable population mean age at net maternity in the ith component, or

$$\kappa_i = \Sigma_j \ j \lambda_i^{-j} \ p_j \ b_j \tag{4.58}$$

To generalize the concept of momentum to component as well as age and time, Schoen and Kim (1992) extended the idea underlying equation (4.49) and wrote

$$\Omega_{ijt} = N_{ijt}/N_{jt} \tag{4.59}$$

It follows that $\Sigma_i \ \Omega_{ijt} = 1$. The discrete form of the $\Omega(x, t)$ of equation (4.49) is thus $\Omega_{1jt}$. However, while $\Omega_{1jt}$ approaches 1 as a population approaches stability, $\Omega_{ijt}(i \neq 1)$ approaches 0 as the subordinate components disappear.

To relate age-specific growth $r_{jt}$ to the roots of the Leslie matrix, equations (4.54) and (4.55) can be combined to write

$$N_{j,t+1} = \Sigma_i \ \lambda_i \ N_{ijt} = \exp[r_{j,t+1}] \ N_{jt} \tag{4.60}$$

Using equation (4.59), equation (4.60) becomes

$$\exp[r_{j,t+1}] = \Sigma_i \ \lambda_i \ \Omega_{ijt} \tag{4.61}$$

Age-specific growth is thus a momentum-weighted average of the growth in each spectral component. As the population approaches stability, $\exp[r_{j,t+1}]$ approaches $\lambda_1$.

Matrix notation allows all ages to be combined in one expression. Let $\gamma$ element column vector $\exp[\mathbf{r}_{t+1}]$ have $\exp[r_{j,t+1}]$ as its jth element, let $\gamma$ element column vector $\lambda$ have $\lambda_i$ as its ith element, and let $\gamma \times \gamma$ matrix $\boldsymbol{\Omega}_t$ have $\Omega_{jit}$ as its (i,j)th element. Then equation (4.61) leads to

$$\exp[\mathbf{r}_{t+1}] = \boldsymbol{\Omega}_t \ \boldsymbol{\lambda} \tag{4.62}$$

As the population approaches stability, $\boldsymbol{\Omega}_t$ becomes a matrix with ones in the first column and zeros elsewhere.

Alternatively, we can relate age-specific growth and the roots of the Leslie matrix through the component-specific reproductive contribution function $q_{ijt}$, given by

$$q_{ijt} = N_{jt}\ v_{ij}/Q_{it} \tag{4.63}$$

where $\Sigma_j\ q_{ijt} = 1$. At time t, $q_{ijt}$ is the contribution of the jth age group to the stable equivalent of the ith component. At time $t+1$

$$\begin{aligned}\Sigma_j\ q_{ij,t+1} &= 1 = \Sigma_j\ N_{j,t+1}\ v_{ij}/Q_{i,t+1}\\ &= \Sigma_j\ \exp[r_{j,t+1}]\ N_{jt}\ v_{ij}/[\lambda_i\ Q_{it}]\end{aligned} \tag{4.64}$$

using equation (4.54) and the fact that $Q_{it}$ grows by factor $\lambda_i$. Rearranging equation (4.64) yields

$$\lambda_i = \Sigma_j\ \exp[r_{j,t+1}]\ q_{ijt} \tag{4.65}$$

which generalizes equation (4.47) to all components. The growth rate of every spectral component is the q-weighted average of the exponentiated age-specific growth rates. Using matrix notation to combine all ages, with $\gamma \times \gamma$ matrix $\mathbf{q}_t$ having (i,j)th element $q_{ijt}$, equation (4.65) yields

$$\exp[\mathbf{r}_{t+1}] = \mathbf{q}_t^{-1}\ \boldsymbol{\lambda} \tag{4.66}$$

As noted by Schoen and Kim (1992), equations (4.62) and (4.66) not only show how the age-specific growth rates are linked to all of the roots of the Leslie matrix, they lead to

$$\boldsymbol{\Omega}_t = \mathbf{q}_t^{-1} \tag{4.67}$$

which shows the intimate relationship between momentum and reproductive contributions across the spectral decomposition. The inverse of the array of reproductive contributions by age to each component gives population momentum, by component, at each age.

## 4.7 SUMMARY

This chapter has examined how demographic behavior influences population size and composition at the margin. In the context of fixed rate models, ordinary and functional differentiation show how demographic measures vary over time. Fertility directly impacts the lower end of the age structure, causing it to pivot around the mean age of the population. In contrast, mortality's influence comes from three distinct and frequently offsetting effects. Introducing the "variable-r" concept allows any population to be expressed in terms of its age-dependent growth, facilitating sensitivity analyses in dynamic populations. Crossover points serve to connect summary measures in populations sharing the same vital rates. An example is Lotka's Ŧ, the crossover age where stable and stationary birth density distributions intersect, which functionally relates the unit birth in a stable population to the $R_0$ births arising from a cohort.

In any population, equation (4.42) shows that change in a broad range of summary measures can be expressed in terms of a weighted covariance between age-specific growth and an age-dependent demographic variable. For example, the change in a crude birth rate is a population weighted covariance between age-specific rates of growth and fertility. Viewing a marginal change as one step in the process of stabilization leads to the fundamental principle that every population is always moving toward the stable equivalent implied by its prevailing rates. The discrete model allows those relationships to be extended to all of the population components implied by the roots of the prevailing population projection matrix. Age-specific growth is a weighted average of component-specific growth rates, where the age-component-specific weights can reflect either momentum or reproductive contributions. Age-dependent growth combines changes over all components, as momentum and reproductive contributions are intimately related across the entire spectrum.

## 4.8 EXERCISES

[See Appendix B for selected answers]

1. Consider stable population characteristic equation (1.29). If there is a neutral change in mortality of the form of equation (4.1), show that the age composition of the stable population remains unchanged. Is there a change in stable r?
2. If mortality changes proportionately as in equation (4.3), verify that equation (4.9) gives a linear approximation to equation (4.4).
3. Consider a stable population with $e(0) = 80$ and no mortality before age 80. Using equation (4.14), what value of r maximizes the fraction of the population between the ages of 15 and 65, and what is that maximum fraction?

4. Under the variable-r approach of Preston and Coale (1982), verify that the characteristic equation can be written

$$1 = \int \exp[-\int_0^x r(a,t)da]\ p(x)\ f(x)\ dx$$

5. Keyfitz (1977: p126) provided a series for the stable mean age at childbearing in terms of the cumulants of cohort net maternity, specifically

$$A^* = \mu - r\sigma^2 + r^2\mu_3/2 - \ldots$$

Use that result to verify the relationship in Kim and Schoen (1993) that

$$\mp = (1/r)\int_0^r A^*(\rho)d\rho$$

where $A^*(\rho)$ is the period mean age at childbearing in a stable population with growth rate $\rho$.
6. Preston, Himes and Eggers (1989) derived the relationship

$$dY_N(t)/dt = 1 - DR[Y_\delta(t) - Y_N(t)] - BR\ Y_N(t)$$

where $Y_\delta(t)$ is the mean age at death in the population and BR and DR are the population's crude rates of birth and death, respectively. Show that relationship is algebraically equivalent to equation (4.44).
7. Using equation (4.54), show that $R_{Nt}$, the overall rate of population growth between times t–1 and t, can be expressed as both (i) the arithmetic mean of the age-specific growth rates weighted by the population at time t–1 and (ii) the harmonic mean of the age-specific growth rates weighted by the population at time t.
8. If $\Omega(t)$ is the momentum at time t in a fixed rate, $r = 0$, population transitioning to stationarity, show that $\partial\Omega(t)/\partial t = -\Omega(t)R_N(t)$.
9. Using equation (4.54), verify the following first differences with respect to time
   a. $\Delta N_j = N_j[\exp(r_j) - 1]$, where $N_j$ is the population in age group j and $r_j$ is the growth rate of that population
   b. $\Delta c_{Nj} = c_{Nj}[\exp(r_j - R_N) - 1]$, where $c_{Nj}$ is the population fraction in age group j.
   c. $\Delta BR = BR[\exp(R_B - R_N) - 1]$
   d. $\Delta DR = DR[\exp(R_D - R_N) - 1]$, where $R_D$ is the rate of increase in deaths
   e. $\Delta Y_N = Y_N[\exp(R_{Nx} - R_N) - 1]$, where $R_{Nx}$ is the rate of increase in Nx (births times age)
   f. $\Delta Y_\delta = Y_\delta[\exp(R_{Dx} - R_D) - 1]$, where $R_{Dx}$ is the rate of increase in Dx (deaths times age)

10. Verify equation (4.67) algebraically, with initial population **x** and PPM **A** given by

$$\mathbf{x} = \begin{bmatrix} 1 \\ x \end{bmatrix} \quad ; \quad \mathbf{A} = \begin{bmatrix} a & 1-a \\ 1 & 0 \end{bmatrix}$$

11. Show that

$$(d/dr)\ \ln\ d = Y_r - Y_\delta$$

where $Y_\delta$ is defined in Exercise 6 and d is the stable population crude death rate (from Preston, Heuveline and Guillot 2001: p158).

# CHAPTER 5

# LONGEVITY AND DYNAMIC MORTALITY

## 5.1 INTRODUCTION

Mortality is the core demographic process that has been studied the longest, with work dating back to Graunt (1662) and the dawn of demography. Methodologically, it is also the most straightforward—everyone dies, and does so exactly once. From a cohort perspective, analyses do not involve issues related to the ultimate level of mortality, but only questions of timing. Mortality does impact rates of net maternity, but at contemporary levels the mortality effect on the size of future cohorts is miniscule. Mortality can thus be isolated from fertility and studied on its own. Those characteristics greatly simplify fixed rate analyses, but changing mortality rates raise issues that have only recently become subjects of research, and dynamic mortality models are just beginning to be developed.

In this chapter, we begin with a brief overview of mortality trends, and identify a new analytical perspective that emerges in dynamic mortality models. After examining period-cohort relationships in the context of changing death rates, we consider dynamic models and the relationships that arise in them.

## 5.2 AN OVERVIEW OF CHANGES IN HUMAN MORTALITY

The 20th century was a time of unprecedented improvements in mortality. In 1900, life expectancy at birth in the healthiest countries was around 40 years; in 2000, it was about 80 years. The Epidemiological Transition (Omran, 1971) experienced by much of the world saw the leading causes of death shift from infectious to degenerative diseases, with the most profound changes among children and young adults. The last decades of the 20th century saw sustained improvements in mortality at advanced ages in most industrial countries (Kannisto et al., 1994), and a rapid growth in the number of centenarians. By the end of the century, mortality declines had replaced fertility declines as the dominant factor producing population aging in the West (Preston, Himes and Eggers 1989).

One striking feature of those epochal changes is that despite very large declines in the level of mortality, the shape of the age curve of death risks changed only modestly. Coale and Demeny (1983, p26) produced model life tables that spanned female life expectancies at birth from 20 to 80 years. Figure 5.1 shows how both model and observed female age-specific probabilities of death follow quite similar age patterns at very different levels of mortality, one level implying a life expectancy at birth of 38.6 years and the other an e(0) of 75.2 years. Table 5.1 summarizes mortality changes using Coale-Demeny "West" model life table values. As life expectancy rises from 20 to 80, the mean age of the stationary population increases, reaching almost 41 years when life expectancy is 80. The standard deviation of the distribution of deaths also changes, but not

*Figure 5.1. Observed and Coale-Demeny Model "West" Life Table Age-Specific Probabilities of Death in a High and a Low Mortality Population*

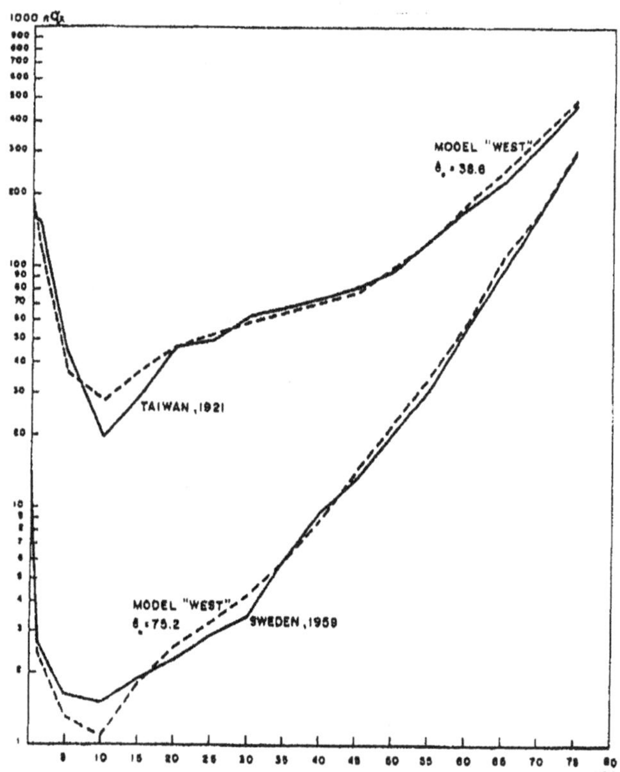


*Source:* Coale and Demeny (1966, p.27). Reprinted with permission from *Regional Model Life Tables and Stable Populations*, Office of Population Research, Princeton University (c) 1966.

*Table 5.1. Summary Measures of Mortality for Life Expectancies of 20 to 80 Years, Coale-Demeny Model "West" Female Life Tables*

| Mortality Level | Life Expectancy[e(0)] | Mean Age of Stationary Population(A) | Standard Deviation of Ages at Death* | Category of Modal Age at Death** |
|---|---|---|---|---|
| 1 | 20 | 25.53 | 24.92 | 0 |
| 3 | 25 | 27.31 | 27.21 | 0 |
| 5 | 30 | 28.89 | 28.87 | 0 |
| 7 | 35 | 30.32 | 29.96 | 0 |
| 9 | 40 | 31.63 | 30.50 | 0 |
| 11 | 45 | 32.85 | 30.52 | 0 |
| 13 | 50 | 33.97 | 29.95 | 0 |
| 15 | 55 | 34.95 | 28.63 | 75–79 |
| 17 | 60 | 35.96 | 26.74 | 75–79 |
| 19 | 65 | 36.96 | 24.08 | 75–79 |
| 21 | 70 | 37.98 | 20.43 | 75–79 |
| 23 | 75 | 39.29 | 16.39 | 80–84 |
| 25 | 80 | 40.91 | 12.07 | 80–84 |

*Notes:*
* Calculated as the positive square root of e(0)[2A – e(0)].
** Using age categories 0, 1–4, 5–9, ..., 90–94, and 95 and over.
*Source:* Ryder (1975) and Coale and Demeny (1983).

monotonically. At low levels of life expectancy, deaths are concentrated at the youngest ages. With an e(0) of 20 years, 53% of all deaths occur before age 5. As mortality improves the dispersion increases. The standard deviation of deaths reaches its maximum when life expectancy is in the low 40s, a point where the median age at death is around 50 and both extremes of the age distribution account for substantial numbers of deaths. The modal age category of deaths remains age 0 through a life expectancy of 50 years; it then jumps to the 75–79 age group. At life expectancies of 50 and higher, deaths become increasingly concentrated at the older ages, and the dispersion of deaths sharply contracts.

The concentration of deaths at older ages, and the continuing mortality declines at those ages, have renewed speculation about the maximum human life span. Currently there is a fairly general sense that, if such a limit exists, we are not yet close to it (Wilmoth, 1997). Most population projections assume that mortality declines will continue for some time (Bongaarts and Bulatao, 2000), despite the spread of AIDS and continuing concerns about epidemics of respiratory diseases. Considerable work has been done in recent years to improve methods for projecting mortality and to estimate the standard errors associated with those projections (Lee and Carter 1992; Li and Lee 2005; Alho and Spencer 2005; Bongaarts 2005). Here we focus on models of changing mortality, and consider a new perspective that arises in dynamic mortality models.

## 5.3 A THIRD ANALYTICAL PERSPECTIVE: THE WEDGE-PERIOD

Traditionally, demographers have emphasized two analytical perspectives: period and cohort. Brouard (1986) was the first to realize that dynamic mortality analysis involves a third perspective. Consider a model population with a unit number of births each year where mortality is free to vary over age and time. At time t, the population at every age is the number from earlier birth cohorts who have survived to that age under the death rates affecting the cohort at each previous time. The time t model population is not the customary model period population (e.g. a period life table population) because it is not derived from a single set of period rates. It is clearly not a cohort, as it is composed of numerous cohorts. Schoen, Jonsson and Tufis (2004) termed it a wedge-period population because, if the age-time-specific mortality rates underlying it are indicated on an age-time (Lexis) surface, they form a triangular or wedge-shaped pattern.

Brouard (1986) called the number of persons in the wedge-period population the "durée de vie moyenne actuelle." Guillot (2003) translated Brouard's term as the "cross-sectional average length of life," and denoted the total size of the time t population by CAL(t). The analysis in Guillot (2003) saw CAL as a mortality index that could usefully be compared to the period life expectancy. When mortality has been declining over time, the period life expectancy based on the death rates of time t, e(0,t), is greater than CAL(t), as CAL(t) shows the effects of the higher mortality of earlier years. Past mortality declines have implications for the future growth of the wedge-period population. If mortality rates become constant at time t levels, the size of the wedge-period population would grow by a ratio of e(0,t)/CAL(t) as the wedge-period population moves from size CAL(t) to e(0,t), the stationary population size and composition.

## 5.4 PERIOD-COHORT CONSIDERATIONS

### 5.4.1 The Period-Cohort Contrast in Mortality

Although period and cohort rates are drawn from the same array, they have different practical and theoretical implications. People live year by year, and data are typically collected and published annually. Period life tables can generally be calculated from data for a very recent year. Cohort life tables require data spanning over a century, while the lifetime experience of recent birth cohorts is still unknown. Substantively, individuals can be influenced by past experiences, and cohorts represent actual people with a shared history. Ryder (1965), in an influential theoretical paper, argued that cohorts are the agents of social change. In mortality analysis, there is evidence that past experiences, especially at early ages, have a substantial effect on later mortality (Elo and Preston, 1992). Moreover, methodological work has found empirical regularities in patterns of cohort mortality (Hobcraft, Menken, and Preston 1982).

### 5.4.2 Average Cohort Life Expectancy

The period life expectancy given by a life table based on the rates of year t, e(0,t), characterizes the experience of a hypothetical (or synthetic) cohort, not any real group of persons (cf. equation (1.11)). When death rates are changing, that measure should not be interpreted as reflecting the longevity of any actual cohort. One way to relate period and cohort behavior is to calculate a period-specific "Average Cohort Life Expectancy" (ACLE). For time t, (Schoen and Canudas-Romo, 2005) proposed the measure

$$\mathrm{ACLE(t)} = \int \mathrm{e_c(0, t-a)}\ell_\mathrm{c}\mathrm{(a, t-a)da/CAL(t)} \qquad (5.1)$$

with

$$\mathrm{CAL(t)} = \int \ell_\mathrm{c}\mathrm{(a, t-a)\ da}$$

$\mathrm{e_c}$ (0, t − a) represents the life expectancy of the cohort born at time t − a, and $\ell_\mathrm{c}$ (a, t − a) represents the number of survivors to age a, from a unit birth cohort, under the death rates experienced by the cohort born at time t − a. By definition, the sum over age of the $\ell_\mathrm{c}$ (a, t − a) is CAL(t). The ACLE measure is thus a weighted average of the life expectancies of the cohorts alive at time t, where the weights are given by the cohort probabilities of survival to each age. Age-specific weights are needed because the few persons who survive to the highest ages should not receive the same emphasis as newborns, and cohort survivorship provides a logical weight for a measure of cohort survival.

To calculate ACLE(t), one must know the mortality risks to which all cohorts alive at time t are subject throughout their lifetimes, information that spans over 200 years and is not available until a century after time t. The need for cohort data is unavoidable, however, because the purpose of the measure is to calculate average cohort life expectancy. Schoen and Canudas-Romo (2005) advanced several arguments regarding those extensive data requirements. First,

ACLE provides a precise, conceptually clear measure for average cohort life expectancy. It is thus a "gold standard" when it can be calculated, e.g. for populations with a long history of mortality data or in the context of models with known values. Second, when it cannot be calculated, it identifies what needs to be estimated, focusing attention on the estimation of future mortality trends and the implications of variations in those estimates.

When mortality is constant over time, $\mathrm{ACLE(t)} = \mathrm{e}(0,\mathrm{t}) = \mathrm{CAL(t)} = \mathrm{e_c}\ (0,\mathrm{t})$. When mortality is steadily declining, $\mathrm{e_c}\ (0,\mathrm{t}) > \mathrm{ACLE(t)} > \mathrm{e}(0,\mathrm{t}) > \mathrm{CAL(t)}$. When mortality is fluctuating, patterns over time can be more complex. Figure 5.2 shows time trajectories for those four measures in the context of a model with one birth per year and mortality changing cyclically over cohorts. At age x and time t, the force of mortality is given by

$$\mu(\mathrm{x},\mathrm{t}) = \exp[\mathrm{A} + (\mathrm{b}+\mathrm{c})\mathrm{x} - \mathrm{ct} - .25\ \sin(2\pi \mathrm{t}/60)] \qquad (5.2)$$

a modified Gompertz pattern. [In the classic Gompertz pattern, $\mu(\mathrm{x}) = \exp(\mathrm{A} + \mathrm{bx})$; see Section 1.9, Exercise 5.] Parameter $\mathrm{A} = -11.215$ sets the level of $\mu(50, 0)$ at 0.002, a reasonable level for a low mortality population. Parameter b, the rate of mortality increase over age, is given its usual value of 0.1. Parameter c, the rate of mortality decrease over time, is taken as 0.008, a decline of less than 1%

*Figure 5.2. Cohort and Period Life Expectancy, ACLE, and CAL for a Sinusoidally Declining Mortality Model with Cohort Cyclical Fluctuations of 60 Years and Parameters A = −11.215, b = 0.1, c = 0.008 and d = 0.25*

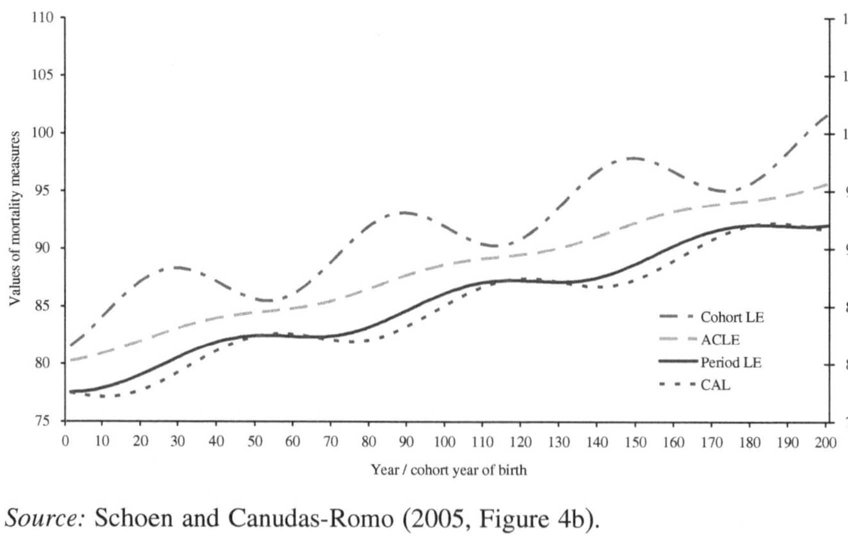


*Source:* Schoen and Canudas-Romo (2005, Figure 4b).

per year. The sine term introduces the fluctuations, which have an amplitude (d) of 0.25 and a cycle length of 60 years. The figure shows that all four measures reflect long term increases over time. Cohort life expectancy is consistently the largest of the four measures, and fluctuates with the greatest amplitude. Period life expectancy and CAL alternate as the smallest measure, with CAL usually lower. Even though CAL is based on a wedge of rates, its fluctuation is about as large as that of period life expectancy. ACLE is the second largest measure and, as a weighted average, shows the underlying trend with the least amount of fluctuation.

## 5.5 A SHIFTING GOMPERTZ MORTALITY CHANGE MODEL

### 5.5.1 The Bongaarts-Feeney Approach

A flexible model of changing mortality was set forth in an important paper by Bongaarts and Feeney (2002). Consider a model population with one birth each year, and recall the $\nu(x, t)$ function of equation (4.21). With N(x,t) the number of persons in the wedge-period population age x at time t, and $N(0, t) = 1$, equation (4.21) implies

$$N(x,t) = \exp[-\int_0^x \nu(a, t)da] \qquad (5.3)$$

a result consistent with equations (4.20) and (4.22). Basically, $\nu(x, t)$ would be a force of mortality if N(x,t) described a life table survivorship curve rather than the number of persons in a population. Bongaarts and Feeney (2002) assumed that v(x,t) followed a Gompertz pattern with

$$v(x,0) = \exp[A + bx] \qquad (5.4)$$

where A is the parameter that specifies the level of $\nu(0, 0)$, and b is the parameter that indicates the pace of increase with age. A Siler model (Siler 1979) can incorporate infant and child mortality, and a logistic model (cf. Bongaarts, 2005) can better capture mortality at high ages, but the Gompertz is by far the most convenient analytical formulation. Bongaarts and Feeney (2002) focus on ages 30 and over to avoid the complications of childhood, adolescent, and young adult mortality. Using equation (5.4) in equation (5.3) and integrating,

$$N(x,0) = \exp[-e^A\{e^{bx} - 1\}/b] \qquad (5.5)$$

Bongaarts and Feeney (2002) introduced change by allowing the v function to shift horizontally over time. Let F(t) be the shift (or offset) in v between times 0 and t, i.e. let

$$v(x,t) = \nu(x - F(t), 0) = \exp[A + b\{x - F(t)\}] \qquad (5.6)$$

*Figure 5.3. Stylized $\mu$ and $\nu$ Curves and the Distances Between Them, Shifting Gompertz Mortality Change Model*

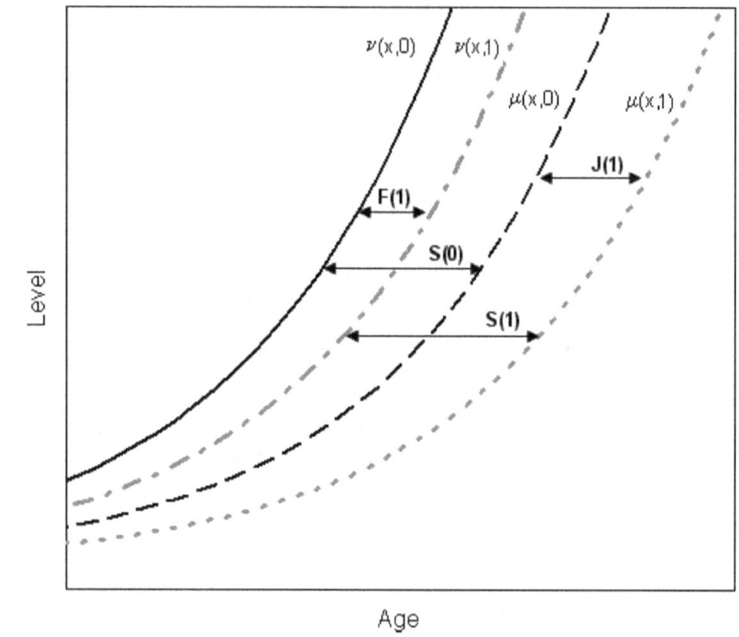


(see Figure 5.3). It follows from equation (5.3) that

$$N(x, t) = \exp[-\int_{F(t)}^{x} \nu(a - F(t)), 0)\, da]$$

$$= \exp[-e^{A}\{e^{b\{x-F(t)\}} - 1\}/b] \qquad (5.7)$$

The lower limit of the integral in equation (5.7) is F(t) because of the shift in the age argument of $\nu$. By definition, whenever the age argument is less than zero, $\nu$ is set equal to zero. Comparing equations (5.5) and (5.7) shows that the N distribution also shifts by F(t) years.

Bongaarts (2002) assumed that death rates also follow a Gompertz pattern, one based on the same parameters as $\nu$. At time 0, let

$$\mu(x, 0) = \exp[A + b\{x - S(0)\}] = \nu(\{x - S(0)\}, 0) \qquad (5.8)$$

In equation (5.8), S(0) years is the offset at time 0 that shifts the $\nu$ function to the $\mu$ function (see Figure 5.3). Hence at time 0, the force of mortality at age x is the same as the $\nu$ function at age $[x - S(0)]$. Since ages less than zero are not defined, $\mu$ is set equal to zero whenever the age argument (here $[x - S(0)]$) is less than zero. At any time t, let S(t) be the offset that shifts $\nu(x, t)$ to $\mu(x, t)$. Hence

$$\mu(x, t) = \nu(\{x - S(t)\}, t) = e^{-bS(t)}\nu(x, t) \tag{5.9}$$

Bongaarts and Feeney (2002) observed that multiplying the force of mortality by a constant at all ages is equivalent (except at the lowest ages) to shifting the $\mu$ curve horizontally. Accordingly, at time $t > 0$ we can write

$$\mu(x, t) = \exp[A + b\{x - S(0) - J(t)\}] \tag{5.10}$$

for $[x - S(0) - J(t)] \geq 0$, where J(t) is a function that indicates the horizontal shift in the force of mortality curve from time 0 to time t (see Figure 5.3). It follows that $\mu(x, t)$ equals $\mu(x, 0)$ times $\exp[-bJ(t)]$, and is the same as $\mu(x - J(t), 0)$. If $J(t) > 0$, the force of mortality curve shifts to the right and mortality declines, while if $J(t) < 0$, the force of mortality curve shifts to the left and mortality rises. When $J(t) > 0$, the size of J(t) equals the rise in period life expectancy from initial conditions, as there is no mortality below age J(t).

Let $\ell_{LT}(x, t)$ be the number of survivors in a life table cohort of one person under the mortality prevailing at time t. Then with mortality given by $\mu(x, t)$, equations (1.5), (5.8), and (5.10) give

$$\ell_{LT}(x, t) = \exp[-\int_{S(0)+J(t)}^{x} \mu(a, t)\, da] = \exp[-e^{A}\{e^{bx-b[S(0)+J(t)]} - 1\}/b] \tag{5.11}$$

where the last equality follows from carrying out the integration. The lower limit in the integral is $S(0) + J(t)$ in order to avoid the "negative ages" where $\mu$ is set to zero. Note that $\ell_{LT}(x, t)$ has the same form as N(x,t) in equation (5.7), except that the shift is $[S(0) + J(t)]$ years instead of F(t) years. Since S(t) is the shift between $\mu(x, t)$ and $\nu(x, t)$, it is also the shift between the $\ell_{LT}$ and N schedules, so we must have

$$e(0,t) = CAL(t) + S(t) \tag{5.12}$$

Recall the age dependent growth rate r(x,t) of equation (4.18). It follows from equations (5.6) and (5.7) that

$$\partial \ln N(x,t)/\partial t = r(x,t) = \nu(x,t)\ dF(t)/dt$$

where the last equality is found by carrying out the partial differentiation. Hence, at every age, r(x,t), the proportional change in N(x,t) over time, equals $\nu(x,t)$, the proportional change in N(x,t) over age, times the change in the offset between N(x,0) and N(x,t). Since $\nu$, $\mu$, and r are related by equation (4.22), we have

$$\mu(x,t) = \nu(x,t)\{1 - dF(t)/dt\} = v(x,t)\{1 - dCAL(t)/dt\} \qquad (5.13)$$

The last equality in equation (5.13) follows from

$$d\ F(t)/dt = d\ CAL(t)/dt \qquad (5.14)$$

which holds because $\nu$ and N shift by the same amount. Equation (5.13) shows that, at a given time, a constant factor transforms the age-specific $\nu$ function into the age-specific $\mu$ function (at all ages above F(t)), confirming the shift in $\mu$. That relationship between $\nu$, $\mu$, and F generally holds in models with shifts (i.e. offsets where "negative ages" have zero rates), as it does not depend on Gompertz mortality. Combining equations (5.9), (5.13), and (5.14), we find

$$d\ CAL(t)/dt = 1 - e^{-bS(t)} = 1 - \mu(x,t)/\nu(x,t) \qquad (5.15)$$

Equation (5.15) is significant as it relates the change in total population size to both the offset between $\mu$ and $\nu$ and the ratio of the $\mu$ and $\nu$ functions.

With a cohort size of one, the number of persons in any life table stationary population is the same as the average age at death in that stationary population. Bongaarts and Feeney (2002) showed that relationship holds in the shifting Gompertz model as well, since

$$\begin{aligned} Y_\delta(t) &= \textstyle\int x\mu(x,t)\ N(x,t)\ dx / \int \mu(x,t)\ N(x,t)\ dx \\ &= \textstyle\int x\ \nu(x,t)\ N(x,t)\ dx / \int \nu(x,t)\ N(x,t)\ dx = CAL(t) \end{aligned} \qquad (5.16)$$

where $Y_\delta(t)$ is the time t wedge-period population average age at death. The first equality in equation (5.16) expresses the definition of $Y_\delta(t)$. The second equality follows from equation (5.13), where the factor of $(1 - dF(t)/dt)$ at all ages drops out. The final result comes from recalling that $\nu$ and N are related to each other in the same way as $\mu$ and $\ell_{LT}$, and for the hypothetical cohort of one person where $\nu(x,t)$ is the force of mortality, the average age at death equals the size of the stationary population, i.e. CAL(t).

Equations (5.12), (5.13), (5.15), and (5.16) provide the central relationships in the Bongaarts-Feeney approach (see the summary in Table 5.2). They reflect a dynamic population where shifting Gompertz mortality implies a shifting wedge-period population. Aside from the youngest ages, the relative age structure of the population remains constant, and the change in population size at any time reflects the ratio between the $\mu$ and $\nu$ functions.

Bongaarts and Feeney (2002) saw CAL(t) as a timing adjusted measure of longevity, an interpretation that has been disputed (Goldstein, 2005; Rodriguez 2006; Wachter, 2005). Since longevity is inherently a measure of the timing of death, it is difficult to conceptually specify what a meaningful timing adjustment should involve (Guillot, 2006). Figure 5.2 indicates that CAL is affected by fluctuations in cohort mortality to about the same extent as period life expectancy. When mortality is monotonically declining, CAL(t) is always less than the life expectancy of year t; the analysis in Wachter (2005) showed that CAL(t) is essentially a weighted average of life expectancies for t and earlier years. It should be noted that the timing adjustment interpretation is not a part of the dynamic model. There is no doubt that the shifting Gompertz model is a significant step forward in dynamic mortality modeling, and that CAL is an important indicator that simultaneously reflects both the total size of the model population and its average age at death.

### 5.5.2 Relationships in a Shifting Gompertz Model

It is worth examining the shifting Gompertz model further to explore some additional relationships among model functions. Equation (5.15) can be rearranged and written as

$$\exp[-bS(t)] = \exp[-b\{S(0) + J(t)\}]\ \exp[bF(t)]$$

from which it follows that

$$S(t) = S(0) + J(t) - F(t) \qquad (5.17)$$

a new result that describes the additive relationship linking the different offsets. Figure 5.3 provides an illustration consistent with equation (5.17).

Equations (5.14) and (5.15) relate the derivative of CAL(t), or F(t), to S(t). It is thus straightforward to go from a known, differentiable F(t) [or $\nu(x, t)$] function, with Gompertz parameters A and b, to S(t), then J(t), and on to the other model functions (see Exercise 7). Typical demographic practice, however, begins with the force of mortality and its pattern of change over time, i.e. with the J(t) function. With J(t) known and S(t) given by equation (5.17), the procedure is not straightforward because differential equations (5.13)–(5.15) are generally not analytically soluble for F(t). As we see below, the reason is that the change in F (or $\nu$) associated with a change in J (or $\mu$) depends in a fairly complicated way on an earlier value of F, or on the difference between $\mu$ and $\nu$ (i.e. on S).

To find an equation for F(t) in terms of J(t), we proceed indirectly and compare N(x,t) from equation (5.7) with an expression for N(x,t) found by applying equation (1.5) and surviving $N(x-1, t-1)$. To facilitate that process, we constrain $\mu(x, t)$ to be the force of mortality at age x from exact time $t-1$ through exact time t, for all x satisfying $[x-1+S(0)+J(t)] > 0$. Then we can determine the population at time t by surviving the population at time $t-1$ using the expression

$$N(x,t) = N(x-1, t-1)\ \exp[-\int_{X-1}^{X} \mu(a, t)\ da] \tag{5.18}$$

To evaluate equation (5.18), note that $N(x-1, t-1)$ can be written in terms of the $\nu(x, t-1)$. Using equation (5.6), we find

$$\begin{aligned} N(x-1, t-1) &= \exp[-\int_{0}^{X-1} \nu(a, t-1)\ da] \\ &= \exp[-\int_{F(t-1)}^{X-1} \nu(a-F(t-1), 0)\ da \\ &= \exp[-e^{A}\{e^{b[x-1-F(t-1)]} - 1\}/b] \end{aligned} \tag{5.19}$$

where the lower limit of integration becomes $F(t-1)$ when the age argument of $\nu$ shifts. The second (exponential) term on the right side of equation (5.18) can be explicitly integrated using equation (5.10). Combining that result with equation (5.19) yields

$$N(x,t) = \exp[-e^{A}\{e^{b(x-1)-bF(t-1)} - 1\}/b - e^{A-bS(0)-bJ(t)}\{e^{bx} - e^{b(x-1)}\}/b] \tag{5.20}$$

Equating the expressions for N(x,t) in equations (5.7) and (5.20), simplifying, and rearranging yields a new expression for F(t), i.e.

$$F(t) = -\ln[e^{-b-bF(t-1)} - e^{-b-b\{S(0)+J(t)\}} + e^{-b\{S(0)+J(t)\}}]/b \tag{5.21}$$

The value of F(t) is independent of age, as the terms in equations (5.7) and (5.20) that involve age drop out. The shift in the $\nu$ (or N) function depends on initial conditions (i.e. the Gompertz parameter b and S(0), the time 0 offset between $\nu$ and $\mu$); on J(t), the overall shift in the force of mortality between times 0 and t; and on F(t − 1) or, using equation (5.17), on S(t − 1) and J(t − 1). Values at time (t − 1) thus exert a considerable, nonlinear influence on the value of F, and hence $\nu$, at time t.

Consider equation (5.21) when $\nu = \mu$, that is when F(t − 1) = S(0) + J(t). In that case, the first two terms of the logarithm in equation (5.21) cancel each other, and we have F(t) = F(t − 1). Hence, in the stationary population case, F(t) remains constant [and $N(x,t) = \ell_{LT}(x,t)$]. Examination of equation (5.21) shows that when J(t) = J(t − 1) but F(t − 1) differs from S(0) + J(t), the value of F(t) is closer to S(0) + J(t) than F(t − 1) is (see Exercise 10). In other words, $\nu$ moves toward $\mu$ when mortality is constant. Since equation (5.14) shows that the derivative of F(t) is the same as the derivative of CAL(t), equation (5.21) indicates that the change in CAL(t) depends on circumstances prior to time t. The shape of the N schedule, at ages over F, is fixed, but the size of the shift in N at any given time depends on the past and generally cannot be determined from a simple relationship.

The shifting Gompertz mortality model involves numerous variables, some of which are not standard. A summary of model variable relationships, including those of this subsection, can be found in Table 5.2.

*Table 5.2. Functions and Relationships in a Shifting Gompertz Mortality Change Model*

| | |
|---|---|
| A. Wedge-Period Functions | |
| $\nu(x,t) = [1/N(x,t)]\{\partial N(x,t)/\partial x\}$ | (4.21) |
| $\nu(x,t) = \nu(x - F(t), 0) = \exp[A + b\{x - F(t)\}]$ | (5.6) |
| $N(x,t) = \exp[-\int \nu(a - F(t)), 0)\, da] = \exp[-e^A\{e^{b\{x-F(t)\}} - 1\}/b]$ | (5.7) |
| B. Life Table Functions | |
| $\mu(x) = [-1/\ell_{LT}(x)]\{d\ell_{LT}(x)/dx\}$ | (1.3) |
| $\mu(x,t) = \mu(x - J(t), 0) = \exp[A + b\{x - S(0) - J(t)\}]$ | (5.10) |
| $\ell_{LT}(x,t) = \exp[-\int_{S(0)+J(t)}^{x} \mu(a,t)\, da] = \exp[-e^A\{e^{bx-b[S(0)+J(t)]} - 1\}/b]$ | (5.11) |
| C. Offset Function and Change Relationships | |
| $\nu(x,t) = \mu(x,t) + r(x,t)$ | (4.22) |
| $\mu(x,t) = e^{-bS(t)}\, \nu(x,t)$ | (5.9) |
| $e(0,t) = CAL(t) + S(t)$ | (5.12) |
| $\mu(x,t) = \nu(x,t)\{1 - dF(t)/dt\} = \nu(x,t)\ \{1 - dCAL(t)/dt\}$ | (5.13) |
| $dF(t)/dt = d\ CAL(t)/dt$ | (5.14) |
| $d\ CAL(t)/dt = 1 - e^{-bS(t)} = 1 - \mu(x,t)/\nu(x,t)$ | (5.15) |
| $CAL(t) = Y_\delta(t)$ | (5.16) |
| $S(t) = S(0) + J(t) - F(t)$ | (5.17) |
| $F(t) = -\ln[e^{-b-bF(t-1)} - e^{-b-b\{S(0)+J(t)\}} + e^{-b\{S(0)+J(t)\}}]/b$ | (5.21) |

## 5.6 MODELS OF CONSTANT MORTALITY DECLINE OVER TIME

### 5.6.1 Three Models With Linearly Changing Mortality

To find more tractable expressions linking shifts in mortality to shifts in population size and structure, we need to specialize the general relationships in Table 5.2 to specific patterns of change. A restrictive yet simple and reasonable kind of mortality change is a linear decline over time. Consider the situation where mortality at age x and time t is given by the Gompertz form relationship

$$\mu(x, t) = \exp[A + bx - ct] \tag{5.22}$$

where parameter c indicates the constant pace of decline in mortality, at all ages, over time. Equation (5.22), the basis of the "Gompertz linear change model", resembles that of a linearly shifting Gompertz model, except that under equation (5.22) the force of mortality at all ages follows an exponential form and does not "slide" at the lowest ages. Under equation (5.22), the model relationships in Table 5.2 hold only approximately. Numerically both models typically yield very similar results, unless c is close to, or greater than, b.

Vaupel (1986) found that equation (5.22) led to an approximately constant change in period life expectancy, i.e.

$$de(0,t)/dt \approx c/b \tag{5.23}$$

Hence if mortality improves by 1% a year ($c = 0.01$) and $b = 0.1$, life expectancy would increase by about 0.1 years each year. Vaupel (1986) also explored the relationship between mortality change and entropy. We previously encountered entropy in Section 2.3, where it referred to the net maternity distribution and was related to the speed of convergence to stability. Here we are interested in the entropy of a survivorship function, which Vaupel (1986) noted can be interpreted as a measure of the heterogeneity of a population with respect to mortality at different ages. If all deaths occur at the same age, entropy (H) is zero. If the force of mortality is constant over age, entropy is one. In the life table context, with a cohort size of 1,

$$H_{LT}(t) = -\int \ell_{LT}(x, t) \ \ln \ \ell_{LT}(x, t) \ dx/e(0, t) \tag{5.24a}$$

In the wedge-period model context, assuming one birth each year, the analogous entropy at time t can be written as

$$H(t) = -\int N(x, t) \ \ln \ N(x, t) \ dx/CAL(t) \tag{5.24b}$$

When mortality follows equation (5.22), Vaupel (1986) gave the approximate relationship

$$H_{LT}(t) \approx 1/\{be(0, t)\} \tag{5.25}$$

indicating that entropy would steadily decline as life expectancy increased.

Schoen, Jonsson and Tufis (2004) further explored the implications of equation (5.22). They found that the increase in total population size is approximately linear, with

$$CAL(t+1) - CAL(t) = 1 - D(t) \approx c/b \tag{5.26}$$

where D(t) represents the total number of deaths in the wedge-period population during year t. In the Gompertz linear change model, CAL(t) and e(0,t) increase at the same pace. The annual number of deaths remains approximately constant, i.e.

$$D(t) \approx (b - c)/b \tag{5.27}$$

Schoen, Jonsson and Tufis (2004) also found close connections to entropy. The wedge-period population crude rate of natural increase is proportional to entropy, with

$$d\ \ln\ CAL(t)/dt = RNI(t) \approx cH(t) \tag{5.28}$$

where RNI(t) is the rate of natural increase at time t. The crude birth rate (CBR) and the crude death rate (CDR) are also proportional to entropy, and at time t can be approximated by

$$\begin{aligned} CBR(t) &\approx b\ H(t) \\ CDR(t) &\approx (b - c)H(t) \end{aligned} \tag{5.29}$$

Comparing the Gompertz linear change model of equation (5.22) to the shifting Gompertz model, Schoen, Jonsson and Tufis (2004) found that under equation (5.22) the wedge-period average age at death is approximately though not exactly equal to CAL. Equation (5.26) suggests that the derivative of CAL is about c/b, which implies that S(t) is roughly $c/b^2$ (from equation (5.15), assuming that $[\ln y] \approx [y - 1]$).

By way of comparison, in a "linearly shifting Gompertz model", we can let $J(t) = (c/b)t$, with $(c/b) < 1$. Equations (5.14) and (5.15) then lead to a constant S(t), i.e.

$$S = -\ln[1 - (c/b)]/b \tag{5.30}$$

Using equation (5.30), equation (5.17) yields $J(t) = F(t) = (c/b)t$. Thus with linear shifts, $\mu$ and $\nu$ (and N) move linearly over time at the same pace.

Goldstein and Wachter (2006) considered an "extended Gompertz linear change" model where mortality follows a linearly changing Gompertz function. Under equation (5.22), survivorship from age 0 to age x at time 0 is given by

$$\ell_{LT}(x, 0) = \exp[-(e^{A+bx} - e^{A})/b] \tag{5.31}$$

In the extended Gompertz model, the survivorship function is instead

$$\ell_{LT}(x, 0) = \exp[-(e^{A+bx})/b] \tag{5.32}$$

in effect considering survivorship from age $(-\infty)$ to age x. Numerically, the difference between equations (5.31) and (5.32) is generally quite small. With $e^{A}$ on the order of 0.00002, the additional exposure in the extended Gompertz linear change model is typically to miniscule forces of mortality. Analytically, equation (5.32) simplifies matters because it eliminates the $e^{A}$ term while preserving the linear shift relationship

$$\mu(x, t) = \mu(x - [c/b]t, 0) \tag{5.33}$$

Goldstein and Wachter (2006) noted that equation (5.32) implies

$$e(0,0) = [(\ln b) - A - \gamma^{E}]/b \tag{5.34}$$

where $\gamma^{E}$ is Euler's constant, approximately 0.577215 (cf. Pollard and Valkovics 1992). As in equation (5.23), life expectancy increases by c/b years per year. The life expectancy of the cohort born in year t is given by

$$\begin{aligned} e_c(0, t) &\approx CAL(t)[b/(b - c)] \\ &\approx e(0, 0) + [c/(b - c)]\{e(0, 0) + t - (1/b)\} \end{aligned} \tag{5.35}$$

In the extended Gompertz linear change model, cohort life expectancy increases linearly over time at a more rapid pace than period life expectancy.

### 5.6.2 Gaps and Lags Between Period and Cohort Life Expectancies

Goldstein and Wachter (2006) pioneered the investigation of gaps and lags in populations with changing mortality. The gap at time t, $\gamma(t)$, is the difference between the life expectancy of the cohort born in year t and period life expectancy at time t, that is

$$\gamma(t) = e_c(0, t) - e(0, t) \quad (5.36)$$

The forward lag in year t, $\lambda_f(t)$, is the number of years before period life expectancy equals $e_c(0, t)$. The backward lag in year t, $\lambda_b(t)$, is the number of years one must move back before the cohort life expectancy of the cohort born in that year equals e(0,t). Figure 5.4 diagrams the relationships between gaps and lags and period and cohort life expectancies.

In the context of a linearly shifting Gompertz model, Goldstein and Wachter (2006) found that the gap can be approximated by

$$\gamma(t) \approx [c/\{b(b-c)\}][ct - A + \ln b - 1.5772] \quad (5.37)$$

indicating linear growth over time at a pace reflecting the speed of the mortality decline.

*Figure 5.4. Gaps and Lags Between Period and Cohort Life Expectancies*

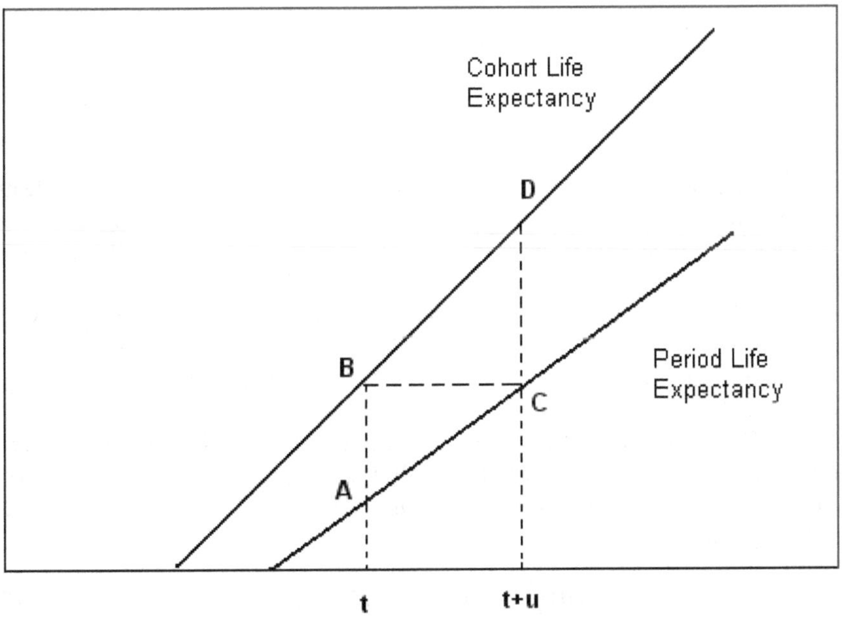


*Note:* Line segment AB is the gap at time t and line segment CD is the gap at time t + u.
Line segment BC is the forward lag at time t and the backward lag at time t + u.

Canudas-Romo and Schoen (2005) found that in a Gompertz linear change model gaps and lags were related by

$$\gamma(t)/\lambda_f(t) \approx c/b$$
$$\gamma(t)/\lambda_b(t) \approx c/(b-c) \qquad (5.38)$$

Numerically, with $b = 0.1$, $c = 0.01$, and $A = -11$, the gap is about 8 years, $\lambda_f$ is about 80 years, and $\lambda_b$ is about 72 years. Schoen and Canudas-Romo (2005) argued that recent increases in the gap between e(0,t) and the life expectancy of the cohort born in year t are attributable to the substantial late 20th century mortality declines at the high ages, and suggested that the divergence between the period and cohort measures will grow as deaths are increasingly concentrated at the high ages.

Following a separate line of inquiry, Goldstein (2006) showed that in a linearly shifting Gompertz model, CAL(t) is the cohort life expectancy of the cohort born CAL(t) years earlier. Mathematically, that can be written

$$e_c(0, t - CAL(t)) = CAL(t) \qquad (5.39)$$

In this context, CAL(t) reflects a unique lag in cohort longevity.

### 5.6.3 The Modal Age At Death

Regularities in the central tendency of the distribution of ages at death can approached using the mode as well as the mean. The modal age at death is simply the age at which the largest number of deaths occurs, and interest in the mode as a measure of mortality goes back to Lexis (1878). Wilmoth and Horiuchi (1999) noted that in a life table context the mode is signaled by an inflection point in the survivorship curve. Canudas-Romo (2005) showed that at modal age M, the force of mortality equals its log derivative with respect to age (see Exercise 8).

Consider a dynamic model with Gompertz mortality over age and any pattern of change over time that affects all ages proportionally. In the life table based on the rates of time t, Canudas-Romo (2005) found that the modal age at death is given by

$$M(t) = \{\ln[b] - \ln[\mu(0,t)]\}/b \qquad (5.40)$$

where b is the Gompertz age parameter. With the number of survivors to age x under the time t life table given by

$$\ell_{LT}(x,t) = \exp[\mu(0,t)\{1 - e^{bx}\}/b]$$

*Figure 5.5. Life Table Survivorship Functions and Distributions of Deaths for a Linearly Declining Gompertz Mortality Model with Parameters A = −11, b = 0.1 and c = 0.01*

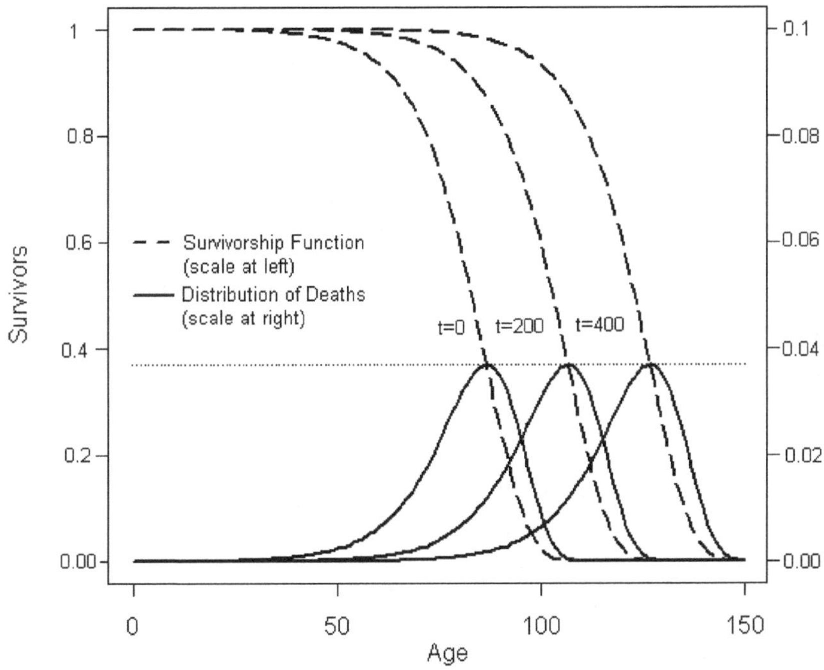


*Note:* The horizontal dashed line indicates $\ell_{LT}(M) = e^{-1} \approx 0.36788$ on the left hand scale and $d_{LT}(M, t) = be^{-1} \approx 0.03679$ on the right hand scale (with $b = 0.1$).
*Source:* Canudas-Romo (2005, Figure 2).

the number of persons at the modal age is

$$\ell_{LT}(M, t) = \exp[\{\mu(0, t)/b\} - 1] \qquad (5.41)$$

and the modal number of life table deaths is

$$d_{LT}(M, t) = \ell_{LT}(M, t)\mu(M, t) = b\exp[\{\mu(0, t)/b\} - 1] \qquad (5.42)$$

as $\mu(M, t) = b$. When $\mu(0, t)$ approaches 0, the number of life table survivors to the modal age approaches

$$\ell_{LT}(M) = e^{-1} \approx .36788 \qquad (5.43)$$

and the modal number of deaths approaches $d_{LT}(M) = b\, e^{-1}$.

Now consider the Gompertz linear change model of equation (5.22). In this case, equation (5.40) yields

$$M(t) = (c/b)t - (A - \ln b)/b \qquad (5.44)$$

If $A = -11$, $b = 0.1$, and $c = 0.01$, $M(t) \approx 86.974 + 0.1t$, and the modal age at death increases by a tenth of a year each year. With those parameters, e(0,t) is approximately 5.27 years less than M(t), as both e(0,t) and M(t) increase at the same pace. Moreover, the distribution of life table deaths around the modal age remains constant in both shape and level, as the mode moves along the age axis at (c/b) years per year (see Figure 5.5). The mode thus specifies a meaningful point in an age distribution of deaths whose form is largely time invariant.

## 5.7 SUMMARY

During the 20th century there were dramatic mortality improvements that greatly reduced death rates at young ages and initiated a period of sustained declines in death rates at older ages. Although mortality has been the subject of demographic analysis for over 300 years, attempts to develop models that explicitly incorporate rates that change over time are relatively recent. A new analytical perspective emerges in such models. It can be termed the wedge-period, and represents a model population based on a constant (unit) annual number of births exposed to risks of death that change over both age and time. Mortality has a cohort component that is not generally captured by period measures, but a measure of average cohort life expectancy, ACLE, provides a way to determine the weighted mean longevity of cohorts active in a given year.

The most general dynamic model currently available is the shifting Gompertz mortality model, where both the age-specific rates and the population structure shift horizontally over time. When mortality changes over time in a linear fashion, the shifting Gompertz model simplifies, and linear relationships emerge to describe changes in life expectancy and in population size and structure. The modal age at death provides a useful summary measure, as that age has a fixed death rate, a fixed survivorship value, and moves linearly with life expectancy. Linear relationships also describe gaps and lags between values of period and cohort longevity. The size of the wedge-period population at time t, CAL(t), is the life expectancy of the cohort born CAL(t) years earlier. Period and cohort measures of longevity are expected to increasingly diverge, as the virtual elimination of childhood mortality makes likely future mortality improvements qualitatively different from past changes.

## 5.8 EXERCISES

[See Appendix B for selected answers]

1. With d(x) denoting the life table distribution of deaths by age (for a unit birth cohort), show that the variance of d(x) is given by $e(0)[2Y_L - e(0)]$, where e(0) is the expectation of life at birth and $Y_L$ is the mean age of the stationary population.
2. Consider a wedge-period population with a history of declining mortality that, at time t, begins a regime of constant mortality at the time t level. Reconcile the result in Guillot (2003, 2005) that momentum ($\Omega$) is given by e(0,t)/CAL(t) with equation (3.1), where $\Omega = b\ e(0)\ Q$ [with b denoting the crude birth rate].
3. Consider a wedge-period population where the Gompertz form force of mortality is given by

$$\mu(x, t) = \exp[A + bx - f(t)]$$

   and let $f(t) = 0$ for $t \leq 0$ and $f(t) = 1$ for $t > 0$. Verify the relationship in Schoen and Canudas-Romo (2005) that

$$CAL(t) = w(t)\ LE(-) + [1 - w(t)]\ LE(+)$$

   where LE(−) and LE(+) are the old and new life expectancies, and w(t) is the time t weight exerted by LE(−).
4. Let d(x,t) be the probability density function describing the time t distribution of deaths, with $\int d(x, t)dx = 1$. In a population with Gompertz mortality, verify the relationship in Vaupel and Canudas-Romo (2003) that

$$1/b \approx \int e(x, t)\ d(x, t)\ dx$$

   where e(x,t) is the expectation of life at age x and time t and b is the Gompertz parameter.
5. Under the Gompertz linear change model of equation (5.22), verify the following results from Schoen, Jonsson and Tufis (2004):
   a. Show that equation (5.23) is an approximation of

$$de(0, t)/dt = (c/b)[1 - \mu(0, t)\ e(0, t)]$$

   and evaluate the size of the additional term.
   b. Show that the Preston-Coale r(x,t) can be approximated by

$$r(x, t) = [c/(b - c)]\mu(x, t)$$

c. Show that with $b = c$,
(i) at every age the mortality of persons born in year t is given by

$$\mu(x, t+x) = \exp[A - ct]$$

(ii) $N(x, t) = \exp[-x\mu(x, t)]$
d. Show that $\ell_{LT}(x, t) = [\ell_{LT}(x, 0)]^{\exp[-ct]}$

6. Consider a Siler model where

$$\mu(x) = A\ \exp[-\alpha x] + B + C\exp[\beta x] \quad , \quad \alpha, \beta > 0.$$

Show that with a unit birth cohort, the number surviving to exact age x is

$$\ell_{LT}(x) = \exp[A(e^{-\alpha x} - 1)/\alpha - Bx + C(1 - e^{\beta x})/\beta]$$

7. Consider a shifting Gompertz model where

$$F(t) = \beta\{1 + \sin[\omega t - \pi/2]\} \quad , |\beta| < 1/2,$$

where angles are measured in radians so $\sin[-\pi/2] = -1$. Show
a. $S(t) = (-1/b)\ \ln\{1 - \beta\omega \cos[\omega t - \pi/2]\}$
b. $J(t) = \beta\{1 + \sin[\omega t - \pi/2]\} - (1/b)\ \ln\{1 - \beta\omega \cos[\omega t - \pi/2]\}$

8. Verify the result in Canudas-Romo (2005) that in a life table population, at modal age at death M, the following relationship always holds:

$$\mu(x) = [1/\mu(x)]\ d\mu(x)/dt$$

9. In a shifting Gompertz model, show that
a. $(\partial/\partial t)\ \ln\ \nu(x, t) = -b(\partial/\partial t)\ F(t)$
b. $(\partial/\partial t)\ \nu(x, t) = -(\partial/\partial x)\ r(x, t)$

10. Let $b = 0.1$ and $[S(0) + J(t)] = 2$. Use equation (5.21) to find F(t) when
a. $F(t-1) = 1.8$
b. $F(t-1) = 1.9$
c. $F(t-1) = 2$
d. $F(t-1) = 2.5$

# CHAPTER 6

# TIMING EFFECTS ON FERTILITY, MARRIAGE, AND DIVORCE

## 6.1 INTRODUCTION

In this chapter we examine fertility, marriage, and divorce, events that may never happen to an individual or that may happen more than once. The focus is on timing effects, in particular how the pace of cohort behavior impacts period measures, with feedback effects generally not considered. The chapter begins with a discussion of period and cohort perspectives on fertility. It then describes two approaches to adjusting period fertility for timing effects, one proposed by Bongaarts and Feeney (1998) and the other involving the Average Cohort Fertility. Those measures are compared in the context of population models and 20th century experience in the United States. Apparently paradoxical period-cohort relationships in fertility are further examined using the dynamic fertility model underlying the Bongaarts-Feeney approach. Extensions of the Average Cohort Fertility approach to first marriage and to divorce are then described and discussed.

## 6.2 THE PERIOD-COHORT CONTRAST

In recent decades, a number of countries have had rates of fertility below the level needed for long term replacement, i.e. below a Net Reproduction Rate (NRR or $R_0$) of 1. At the same time, the mean age at childbearing in many of those countries has been increasing. The combination of those two factors has renewed debate on the interpretation of period fertility measures and on the importance of changes in fertility timing. This chapter examines those issues in depth, drawing heavily on Schoen (2004).

The standard measure of fertility is the Total Fertility Rate (TFR). The TFR can be defined as the sum of the age-specific (female) fertility rates for all ages, and interpreted as the number of children the average woman would have under that specified schedule of fertility (in the absence of mortality and migration). If the birth rates of a particular year (or period) are summed, one gets a period TFR. The period TFR relates to a hypothetical (or synthetic) cohort, and does not describe the experience of any actual group of persons. A cohort TFR is found if the birth rates used follow the experience of an actual birth cohort.

A number of demographers, Norman Ryder (1969; 1980; 1986) foremost among them, have argued that the cohort perspective affords the best way to analyze fertility. Birth cohort measures describe the experience of real groups of women, exposed to the same historical events at the same point in their lives, who share a common reproductive past. Populations have fertility each year, but women have children, one at a time. Fertility theories generally seek to explain completed family size, i.e. cohort fertility. Over time, the cohort TFR (CTFR) has fluctuated less than the period TFR (PTFR), indicating that it is a more stable measure. To cohort advocates, a crucial limitation of the period TFR is that it does not distinguish between a change in the timing (or tempo) of cohort fertility and a change in the level (or quantum) of cohort fertility. For example, a decline in the PTFR of a particular year might mean that the cohorts active in that year are having fewer children, or simply that they have postponed childbearing without changing their ultimate completed family size.

The primacy of the cohort perspective has been undermined by challenges to both its conceptual basis and empirical support. Ni Bhrolchain (1992) saw "period as paramount", noting that the cohort perspective implies a fixed-target view of decision making that is not consistent with observed behavior. Statistical analyses of fertility behavior have repeatedly shown that periods, not cohorts, account for most of the variation in fertility behavior. Characteristic age patterns have been observed in period fertility but, unlike the case in mortality, no such patterns have been found in cohort fertility. The typical pattern of change has been a rise (or a fall) in period fertility at all ages.

Those facts clearly demonstrate the importance of the period perspective, but they can be interpreted in ways much less critical of the cohort view. To say that cohort changes account for less variability in fertility just puts a different slant on the argument that the cohort TFR is a more stable measure. It is the meaningfulness of the measure, not its variability, that counts. What matters theoretically, and in many cases substantively, is completed family size. Women can achieve their long term childbearing goals in many ways, and one should expect that period circumstances will influence fertility timing. The process can be seen as analogous to a drive from one location to another. The driver proceeds toward a given destination, but at different speeds under different traffic conditions.

The argument that the cohort view embodies a fixed reproductive target is a serious criticism, because there is considerable evidence that there is no such target (cf. Lee, 1980). Yet the fact that women change their reproductive goals over time is not problematic unless an extreme cohort position is taken. Returning to the driving analogy, the initial destination can remain meaningful even if traffic conditions cause the driver to stop earlier or go further than first planned. Moreover, the final destination is often the fact of greatest importance.

In sum, both the cohort and period perspectives are valuable in the study of fertility. To ignore either one can be unwise (see Section 6.7. below). Furthermore, given the present focus on timing, we need to compare and contrast period and cohort fertility behavior. The period perspective by itself is not enough because, as Bongaarts and Feeney (1998: 178) recognized, "a notion of 'deferring' or 'advancing' births necessarily refers at some level to cohorts."

## 6.3 THE BONGAARTS-FEENEY APPROACH

In an influential article, Bongaarts and Feeney (1998) proposed a procedure for adjusting the period TFR to eliminate "distortions" produced by changes in the timing of reproduction. Let f(x,t) be the fertility rate for women age x at time t. The TFR for year t can then be written

$$\mathrm{TFR}(t) = \Sigma_x\, f(x,t) \tag{6.1}$$

where the sum ranges over all reproductive ages. Bongaarts and Feeney (1998) defined a year t Total Fertility Rate for births of order o, $\mathrm{TFR}_o(t)$, as the sum of "incidence" rates of order o. Those incidence rates are defined as births of order o to women age x at time t divided by the number of women age x at time t in all birth orders. Consequently

$$\mathrm{TFR}(t) = \Sigma_o\, \mathrm{TFR}_o(t) \tag{6.2}$$

where the sum ranges over all birth orders. The basic measure of timing used is the MAC, the mean age at childbearing, calculated as the unweighted average of the age-specific fertility rates. For each year, an order-specific $\mathrm{MAC}_o(t)$ is found from the incidence rates for each birth order. The change in the timing of fertility at order o during year t, $r_o(t)$, is then given by

$$r_o(t) = [\mathrm{MAC}_o(t+1) - \mathrm{MAC}_o(t-1)]/2 \tag{6.3}$$

Bongaarts and Feeney (1998) advanced a model where the age curve of fertility maintains a fixed shape over time, but can shift along the age axis. For example, assume that at the beginning of year t the reproductive period starts at age 15 and ends at age 46. If the MAC increases by 0.1 years over the course of year t, the reproductive period would shift so that at the end of year t it would begin at 15.1 years and end at 46.1 years. The shape of the age curve of fertility, however, would remain the same. Given that model, the paper proved a new relationship between period and cohort fertility when there was a constant linear change over time in the MAC, and argued that new relationship could also be applied to cases where shifts in the fertility schedule varied over time. Under the Bongaarts-Feeney approach, the order-specific tempo adjusted $TFR_o$ for year t, $TFR_o^*(t)$, is given by

$$TFR_o^*(t) = TFR_o(t)/[1 - r_o(t)] \quad (6.4)$$

The Bongaarts-Feeney tempo adjusted TFR* for all birth orders combined is then found from

$$TFR^*(t) = \Sigma_o TFR_o^*(t) \quad (6.5)$$

When the MAC is rising, the tempo-adjusted TFR* can be considerably larger than the conventional period TFR. Since increases in the MAC characterized many nations during the last quarter of the 20th century, Bongaarts (1998; 1999; 2002) argued that many recent period TFRs in the United States, other developed countries, and a number of developing countries were substantially depressed by timing effects produced by the postponement of births. Bongaarts (1998: p240) predicted that "Once women stop deferring births, the distortion disappears and the very low fertility rates observed in the developed world should rise closer to the two children most couples want."

Other observers disagreed. Lesthaeghe and Willems (1999), Van Imhoff and Keilman (2000), and Frejka and Calot (2001) examined fertility patterns in many low fertility countries and did not see that any increases in fertility were likely in the relatively near term. Van Imhoff and Keilman (2001) and Kohler and Ortega (2002) questioned the use of incidence rather than occurence/exposure rates. In addition, Kim and Schoen (1999; 2000) criticized the methodology underlying the Bongaarts-Feeney adjustment, arguing that when the change in MAC was not linear, the adjustment could be unstable and inaccurate.

Zeng and Land (2002: p270) pointed out that the Bongaarts-Feeney tempo adjustment provides "the average total number of births per woman of a hypothetical cohort that has gone through the imagined extended period with changing tempo but constant quantum and invariant shape of the [fertility] schedule." The usefulness of the Bongaarts-Feeney procedure really hangs on whether an adjusted measure of that kind appropriately captures tempo effects. The classic timing question involves how changes in cohort timing impact period quantum. Yet Bongaarts and Feeney (1998: p275) explicitly assumed that there are no cohort fertility effects, only age, period, parity, and duration since last birth effects. If the period MAC increases, the Bongaarts-Feeney fertility adjustment must be upward because women are exposed to the given fertility rates over a longer reproductive interval. In actuality, the period MAC could increase either because of a change in cohort tempo or a change in cohort quantum; the latter should not prompt a tempo adjustment. The assumption that an increase in the MAC observed during a single year persists through every year of the hypothetical cohort's reproductive life is of particular concern. Because it perpetuates a one time change, the Bongaarts-Feeney adjustment rests on a strong and generally counterfactual assumption.

The combination of unprecedented low levels of fertility and sustained increases in mean ages of childbearing does raise a substantial likelihood that cohort tempo may be impacting the level of period fertility. Given the concerns about the Bongaarts-Feeney adjustment, we need to consider alternative approaches to measuring tempo effects.

## 6.4 AVERAGE COHORT FERTILITY

We consider timing effects to be changes in the level of period fertility that do not reflect changes in completed family size. Other analyses (e.g. Lutz, O'Neill and Scherbov 2003) have included broader effects, such as the slower population growth and age compositional changes associated with a higher MAC, but they introduce additional considerations and are not pursued here. To operationalize our more focused concept of timing, we employ a measure that was independently and roughly simultaneously developed by Butz and Ward (1979) and Ryder (1980). Butz and Ward (1979) called the measure the Timing Index, and it examines the proportions of cohort fertility arising in a particular period.

The cohort TFR for women born in year $\tau$ can be expressed in a form similar to equation (6.1) by

$$\mathrm{CTFR}(\tau) = \Sigma_x \, f(x, \tau + x) \qquad (6.6)$$

The proportion of the CTFR for the cohort born in year $\tau$ that is contributed by women age x (during year $\tau + x$) can be denoted by $\beta(x, \tau + x)$ and expressed as

$$\beta(x, \tau + x) = f(x, \tau + x)/\mathrm{CTFR}(\tau) \qquad (6.7)$$

The Timing Index for year t, TI(t), can then be written

$$\mathrm{TI}(t) = \Sigma_x \, \beta(x, t) \qquad (6.8)$$

In effect, the TI(t) gauges the extent to which the *cohort* fertility of women childbearing during year t occurs in that *period*. There is no timing effect when $\mathrm{TI}(t) = 1$, as the childbearing cohorts have fractions of their lifetime fertility during year t that are consistent with constant cohort timing. When $\mathrm{TI}(t) > 1$, year t has a disproportionately large amount of the cohort fertility of the reproductively active cohorts, indicating that cohort tempo changes favored that year. Analogously, if TI(t) < 1, year t has a disproportionately small amount of the fertility of the reproducing cohorts, indicating that cohort tempo changes avoided that year. The Timing Index thus directly measures the impact of cohort tempo on period quantum.

The Timing Index allows the period TFR to be decomposed into quantum and tempo components. Let the Average Cohort Fertility at time t, ACF(t), be the quantum component. Following Butz and Ward (1979), we have

$$\mathrm{ACF(t)} = \mathrm{TFR(t)}/\mathrm{TI(t)} \tag{6.9}$$

Equation (6.9) has a form similar to that of equation (6.4), the Bongaarts-Feeney adjustment equation, with TI(t) the new tempo component. However, the ACF is not simply an adjusted TFR. Butz and Ward (1979) showed that ACF(t) is a weighted average of the CTFRs of the active cohorts, where the weight at each age is $\beta(\mathrm{x}, \mathrm{t})$. The ACF(t) is not a weighted average of period TFRs and does not represent the fertility of any actual cohort, but provides an average of the CTFRs of the cohorts reproducing at time t.

The ACF has two other features of note. First, it is a strictly behavioral measure of age-specific fertility behavior and its distribution over time. It conveys no information about fertility intentions at any time or about desired family sizes.

Second, the ACF requires information on the lifetime fertility behavior of all actively reproducing cohorts. That information is not available for some 30–35 years after time t, the usual situation encountered in cohort fertility analyses. The problem is unavoidable, since one must know the full age distribution of cohort fertility in order to understand cohort tempo. In historical data, and in population models, those data demands may be met for the times of interest. Moreover, at time t, ACF(t) can be estimated on the basis of assumed patterns of future fertility. The range of ACF(t) values under alternative assumptions can be very informative, and the process serves to illuminate the uncertainties involved in measuring current timing effects.

## 6.5 COMPARING THE BONGAARTS-FEENEY AND ACF APPROACHES

Schoen (2004) compared TFR* and ACF using both model populations and actual data. Seeing how the measures perform in the context of a model where everything is known and manipulable can be revealing, while a comparison using observed rates can show the magnitude of historical timing effects on fertility.

### 6.5.1 Timing Effects in Selected Models

We begin by considering cyclical changes in period tempo with period quantum fixed at 1. Schoen (2004) used models whose age pattern of fertility was that of United States Females, 1975, and introduced a cyclical pattern of change beginning at time 50. To produce period tempo changes, the base age pattern of fertility, f(x) where $\Sigma_x f(x) = 1$, was changed so that for $t > 50$

$$\phi(x, t) = f(x)[1+st]^x/\Sigma_i\{f(i)[1+st]^i\} \qquad (6.10)$$

With $s > 0$, multiplication by $[1+st]^x$ increases fertility rates more at older ages than at younger ages. Dividing each $f(x)[1+st]^x$ value by their sum over age insures that the adjusted period fertility rates at every time sum to 1, while preserving the later age pattern of fertility. That transformation is a convenient tool that has useful analytical properties and yields a reasonable fertility pattern (Coale and Trussell 1974; Schoen and Kim 1996, see Exercise 2).

Figure 6.1 compares the Bongaarts-Feeney TFR* with ACF when s, the annual rate of increase in the fertility transformation, is 0.04 and the length of each cycle is 40 years. The mean age at childbearing, divided by 25 to put it on a comparable scale, is also shown. The ACF cycles moderately with the MAC, while the TFR* cycles with an exaggerated amplitude. During each cycle, there are times (e.g. between years 75 and 80) when the TFR* indicates that period fertility is in one direction from the base value of 1 while the ACF shows that the average fertility of the active cohorts lies in the other direction.

Schoen (2004) also examined models where the MAC rose indefinitely and where the MAC rose for 40 years and then became constant at that higher level. In both cases, the TFR* showed steeper and larger changes than the ACF. Earlier, Kim and Schoen (2000) compared the two measures in a cyclical model with uniform fertility over age. With the cycle length set equal to the length of the reproductive period, both the PTFR and the CTFR were constant at a value of 1. The ACF was also constant at a value of 1, while the TFR* fluctuated greatly. When MAC varied by ±4 years over a cycle of 30 years, the TFR* reached values over 6. In sum, model comparisons indicate that TFR* can misconstrue and exaggerate, sometimes dramatically, the extent of changes in average cohort fertility produced by demographically plausible fluctuations in fertility timing.

### 6.5.2 Timing Effects in the United States, 1917–97

Schoen (2004) compared the ACF and TFR* using published data for the twentieth century United States. Single year period data were used with

*Figure 6.1. Values of the Mean Age at Childbearing (MAC), Bongaarts-Feeney Adjusted Fertility* (TFR*), *and Average Cohort Fertility (ACF) in a Model Population with a Constant Period TFR of 1 that Experiences Cyclical Changes in the Timing of Period Fertility Beginning at Time 50*

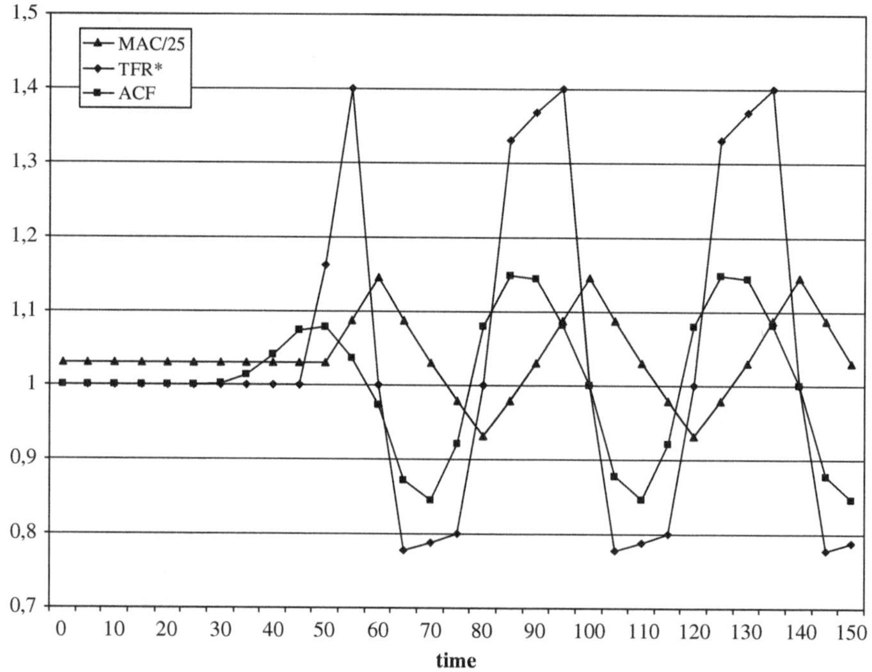


*Note:* Cycles shown have length 40 years with s = 0.04 in the underlying fertility transformation.
*Source:* Schoen (2004, Figure 3d).

single year of age fertility data for ages 15 through 49 and parities 0 through 7 and 8+. Period TFRs were found using equation (6.1), cohort TFRs from equation (6.6), and TFR* from equations (6.2) through (6.5). The ACF was calculated from equations (6.7) through (6.9), but for 1950 and earlier years, when estimates were needed for years before 1917, the ACF was taken from Ryder (1980). To complete the experience of cohorts active after 2001, the latest data year, the average of the (fairly constant) rates for the 1997–2001 period was used. Those estimates affected ACF and CTFR values for cohorts born after 1952, with cohorts born in 1968 and later years having rates estimated for ages under 35.

Figure 6.2 shows values for the period TFR, ACF, and TFR*, as well as the CTFR for the cohort born 26 years earlier. American fertility fluctuated substantially over the twentieth century. The PTFR declined to a low of 2.17 in 1933, rose irregularly to a high of 3.68 in 1957, and then fell to a low of 1.74 in 1976. It remained below 2.00 until 1989, and stayed in the 2.0 to 2.1 range through 2001, the latest data year. The shifted CTFR and the ACF followed similar paths, but with smaller fluctuations. As it represents a weighted average of the completed family size of many different cohorts, the ACF is more stable than the shifted CTFR. In contrast, the TFR* displayed a trajectory similar to the PTFR, frequently with leads or lags of several years, and displayed considerable variability.

Timing effects, as indicated by the difference between the ACF and the PTFR, have often been sizeable. The biggest differences were during the Baby Boom years of 1951–64, where they were as large as two-thirds of a child. Sizeable timing effects also occurred during the 1920s, 1930s, and 1970s. Those of the 1920–27 period have received far less attention than those of the Depression years of the 1930s or the Baby Boom of the 1950s. In the 1920s,

*Figure 6.2. Values of the Period Total Fertility Rate (PTFR), the Average Cohort Fertility (ACF), the Cohort Total Fertility Rate for the Cohort Born in the Year t-26, and the Bongaarts-Feeney Adjusted Measure* (TFR*), *United States, 1917–2001*

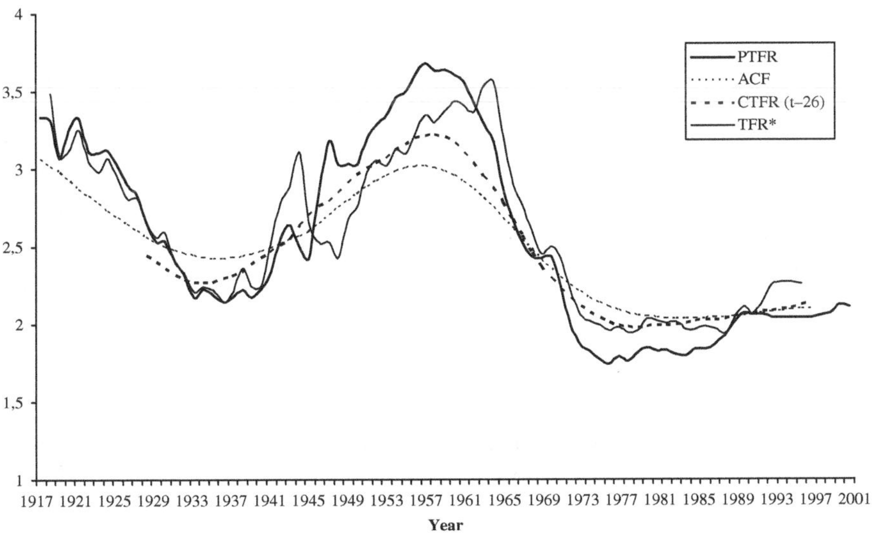


*Source:* Schoen (2004, Figure 4).

timing effects raised the PTFR above the ACF by as much as 0.39 (in 1924), although both measures were generally declining and the MAC changed little. The apparent "acceleration" of fertility in the 1920s is better viewed in the light of the low fertility levels that prevailed during the post-1929 Depression years (Butz and Ward, 1979). As previously noted, the TI and ACF specify timing effects solely with respect to the distribution of cohort fertility over time, and do not imply any intentionality.

Timing effects played a very large and to date underappreciated role in the "Birth Dearth" of the 1970s. Although the ACF was never less than 2.06 during the 1970s, the PTFR fell to a low of 1.74 in 1976. In that year, the ACF was 2.11, indicating that timing effects lowered the PTFR by 0.37, moving the fertility level from above replacement to clearly below replacement. Through 1997, the U.S. has never had an ACF below 2.03, a level reached during 1982–85.

In contrast to the stability shown by the ACF, the TFR* varied nearly as much as the PTFR, sometimes overadjusting, frequently underadjusting and, during the years 1963–66, adjusting in the wrong direction. As in the cyclical model of Figure 6.1, the TFR* showed later maxima and, during the Birth Dearth, an earlier minimum than the ACF. During the 1970s, the TFR* substantially understated the size of the timing effect that produced the lowest PTFRs in U.S. history. When the PTFR was at its minimum of 1.74 in 1976, the TFR* stood at 1.96.

In summary, both model populations and the twentieth century experience of the United States demonstrate that timing effects can substantially raise or lower the period TFR. Measuring timing effects needs to become a routine part of fertility analysis. The ACF provides a way to do so that rests on a solid conceptual foundation, is easily interpreted, and has been validated empirically. The TFR* is less suitable, as it is based on strong and generally unrealistic assumptions and can be unstable empirically. For recent years, when complete data on cohorts are not available, the ACF can be approximated from assumptions on the future course of fertility, a procedure that reflects the uncertainties inherent in measuring the extent of timing effects on current period fertility.

## 6.6 EXAMINING PARITY-SPECIFIC PERIOD TFRs

An alternative approach to achieving a better understanding of fertility behavior is to improve period measures. Ni Bhrolchain (1992) recommended the use of period fertility rates specific not only to age but also to parity and duration since last birth. Fertility is known to vary substantially by parity and duration, and Ryder (1986) concluded that the U.S. Baby Boom was largely the result of higher fertility among women of parities 1 and 2.

Schoen (2006) employed rates prepared by the National Center for Health Statistics for the United States, 1917–97, to analyze the implications of using age-parity-specific fertility rates. Data were not available to calculate fertility rates by duration since last birth. The published fertility rates were used to construct parity status life tables (PSLTs) that, at every age, recognized parities 0 through 7 and 8 or more. Such a PSLT shows the experience of a real or hypothetical cohort of women, born at parity 0, as they move into parities 1 through 8 subject to a given regime of age-parity-specific fertility rates (and no mortality). Each life table cohort was followed, by single years of age, to age 50, and the parity distribution at age 50 furnished each cohort's completed fertility. For example, the proportion of women parity 0 at age 50 provided the proportion childless. With $\ell_p(x)$ the number of women of parity p at exact age x, the PSLT TFR (or PSTFR) is given by

$$\mathrm{PSTFR} = [\Sigma_p p \ \ell_p(50)]/\ell_0(0) \qquad (6.11)$$

where births of parity 9 and higher are ignored.

Schoen (2006) compared the conventional period TFR with the PSTFR over the 1917–97 interval. Because the PSTFR incorporates information on parity, it is based on a different set of population exposures to risk, and generally differs from the period TFR. For the twentieth century United States, the two measures followed a similar pattern, though the PSTFR fluctuated more, having both lower minimum and higher maximum values. The largest difference, 0.55, was in 1957, the peak of the Baby Boom, when the TFR was 3.68 and the PSTFR was 4.23. That substantial gap occurred because high period fertility at young ages and low parities accelerated the movement of the PSLT cohort into higher parities, where they had more years of exposure to the unusually high rates prevailing at those parities. In other words, timing effects on parity as well as age elevated the period PSTFR above the period TFR. Greater specificity in the period fertility rates produced an increase, not a decrease, in the influence of cohort tempo on period quantum.

Schoen (2006) found similar effects with respect to the ultimate parity distribution. Since the 1970s, those distributions have been fairly stable, with about one-third of women having 2 children, and about one-sixth of women ending at each of parities 0, 1, 3, and 4 or more. However, during the years 1951 through 1963, over 30% of the women in the PSLT cohorts had 5 or more children; over 40% had 5 or more children in the PSLTs for 1956 and 1957. No cohort PSLT relating to those years ever reflected a fraction of women having 5 or more children as high as 30%. Although adding parity to age does reveal more of the implications of observed fertility rates, age-parity-specific measures can amplify timing effects.

## 6.7 A DIMINISHING POPULATION WHOSE EVERY COHORT MORE THAN REPLACES ITSELF

In a linearly shifting fertility model of the kind analyzed by Bongaarts and Feeney (1998, see Section 6.3) period and cohort TFRs are constant over time and are related by

$$\text{CTFR} = \text{PTFR}/(1 - \text{r}) \qquad (6.12)$$

where r is the constant annual increase in the mean age at childbearing. Schoen and Jonsson (2003a) noted that an apparent paradox can arise in that model when the MAC is rising. The following discussion draws largely on that work.

Let us consider only female births and assume that there is no mortality below the highest age of childbearing. Equation (6.12) then applies to the period and cohort NRRs. When $r > 0$, it is possible for the period NRR to be less than 1 while the cohort NRR is greater than 1. For example, if the period NRR is 0.9 and $r = 0.2$, then the cohort NRR is $0.9/0.8 = 1.125$. In this case, we either have a population that is growing in size even though every period NRR is below replacement, or a population that is diminishing is size even though every cohort more than replaces itself.

To analyze the birth trajectory in such a model, let us begin with a "rectangular" population, i.e. a population that has one person at every age through the highest age of childbearing. With a period NRR of 1, that population would be stationary. Now let fertility decline so that the period NRR is 0.9, while the fixed-shape fertility schedule begins to shift to higher ages at a rate of 0.2 years per year. In the following year, there would be 0.9 births. Birth cohorts would remain that size until enough time passes for those smaller cohorts to reach reproductive age, after which birth cohort size would fall further below replacement. Starting with a non-stationary age distribution could make the birth trajectory more complicated, but the same result would eventually arise—birth cohort size would decline over time.

Figure 6.3 shows birth trajectories for 6 model populations representing combinations of NRRs of 0.990 and 0.975 with r values of 0, 0.03, and 0.05. With $r > 0$, all of those combinations yield cohort NRRs greater than 1. Figure 6.3 begins at time 150 when initial population compositional effects have largely disappeared, but all trajectories started with a rectangular population at time 0. It is clear that in all 6 cases the annual number of births declines over time. A smaller NRR and a smaller r both decrease the number of births and increase the rate of decline in the number of births, though declines in the NRR have a greater effect.

Models with changing rates expand the scope of demographic analysis and encompass situations that cannot arise in fixed rate models. Here, the

*Figure 6.3. Annual Number of Births in Linearly Shifting Fertility Models by NRR and Annual Change in the Mean Age of Childbearing*

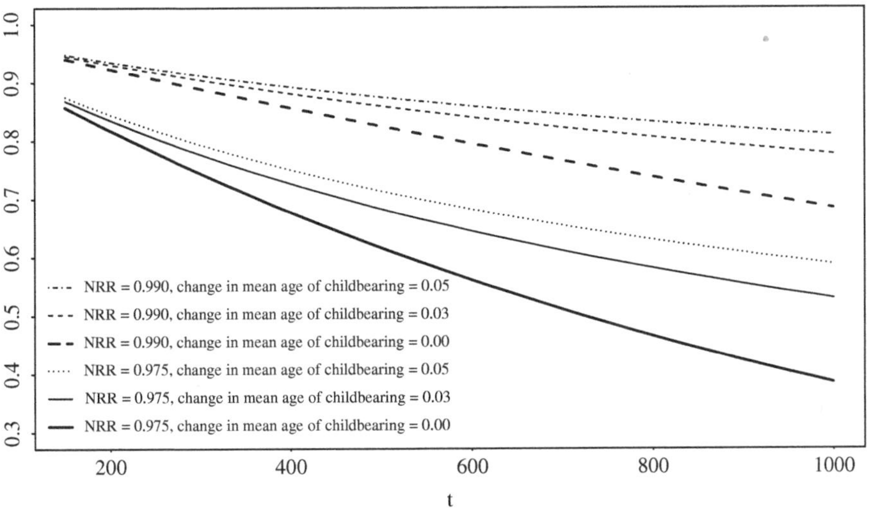


*Source:* Schoen and Jonsson (2003).

apparent paradox of a declining population with cohort NRRs greater than 1 is resolved by realizing that the cohorts that more than replace themselves are continually decreasing in size. With period fertility always below replacement, birth cohort size must eventually decline; timing effects can slow but not stop that process. Period, not cohort, fertility is the key determinant of birth trajectories. Cohort size is a variable that is usually neglected in cohort fertility analyses, but it is worth remembering that cohort size arises from the combination of period fertility and period composition.

## 6.8 TIMING EFFECTS ON FIRST MARRIAGE

The ability of cohort tempo to affect period quantum extends to demographic behaviors other than fertility. Marriage is a prime example. Over the past several decades, many Western countries have experienced large declines in levels of first marriage, accompanied by substantial increases in the average age at first marriage. As in the case of fertility, that raises the question of whether recent period values reflect underlying cohort behavior.

Winkler-Dworak and Engelhardt (2004) analyzed data for Austria, Germany, and Switzerland, and argued that a significant part of the decline in first marriage levels was due to tempo effects. Goldstein (2004) analyzed French data, and came to a similar conclusion. Both studies were based on adaptations of the Bongaarts-Feeney fertility adjustment, however, which reduces confidence in their findings. Here, we focus on the work of Schoen and Canudas-Romo (2005a), which adapted the ACF approach to first marriage behavior and analyzed patterns in England and Wales and the United States.

### 6.8.1 Measuring Timing Effects on First Marriage

Analytically, first marriage differs from fertility because individuals can only enter a first marriage once. The standard measure of the level of first marriage is the proportion ever marrying (PEM). Let m(x,t) be the occurrence/exposure first marriage rate at age x and time t, that is the number of first marriages to persons age x at time t divided by the number of never married persons age x at time t. The time t total first nuptiality ratio, TFNR(t) can then be defined by

$$\mathrm{TFNR}(t) = \Sigma_x\, m(x,t) \qquad (6.13)$$

where the sum ranges over all first marriage ages. The proportion ever marrying under the rates of time t is then given by the usual life table relationship

$$\mathrm{PEM}(t) = 1 - e^{-\mathrm{TFNR}(t)} = 1 - \exp[-\Sigma_x\, m(x,t)] \qquad (6.14)$$

which assumes that the rates are constant within the one-year age intervals. Similar relationships hold for cohorts. For the cohort born at time t, the cohort TFNR is given by

$$\mathrm{CTFNR}(t) = \Sigma_x\, m(x, t+x) \qquad (6.15)$$

and that cohort's proportion ever marrying is

$$\mathrm{CPEM}(t) = 1 - e^{-\mathrm{CTFNR}(t)} = 1 - \exp[-\Sigma_x\, m(x, t+x)] \qquad (6.16)$$

To separate quantum and tempo effects, we follow the logic underlying the TI and ACF. Let a timing effect be a level change in a period PEM that does not reflect a level change in a cohort PEM. For example, if the initial first marriage rate at age x increases by $\Delta m$ at time t and the initial first marriage rate at age $x+n$ and time $t+n$ decreases by $\Delta m$, then the period PEMs at times t and $t+n$ change while the cohort PEM for the cohort born at time $t-x$ remains constant.

For the cohort born at time $\tau$, denote the proportion of the cohort TFNR arising at age x (during year $\tau+x$) as $\beta(x, \tau+x)$. Then

$$\beta(x, \tau+x) = m(x, \tau+x)/\Sigma_a\, m(a, \tau+a)$$
$$= m(x, \tau+x)/CTFNR(\tau) \qquad (6.17)$$

As in equation (6.8), the Timing Index for year t is given by

$$TI(t) = \Sigma_x\, \beta(x, t) \qquad (6.18)$$

Paralleling the case in fertility, TI(t) measures the extent to which year t has a disproportionate share of the first marriage rates of the active cohorts.

The Timing Index gives the tempo component of the TFNR. To obtain the quantum component, which can be termed the adjusted first marriage ratio (AFMR), we proceed in a manner analogous to equation (6.9) and write

$$AFMR(t) = TFNR(t)/TI(t) \qquad (6.19)$$

When $TI(t) > 1$, indicating a concentration of first marriage behavior in year t, equation (6.19) reduces the period TFNR proportionately. When $TI(t) < 1$, indicating an avoidance of year t, equation (6.19) proportionately increases the period TFNR. The time t adjusted proportion ever marrying, $PEM^*(t)$, is then given by

$$PEM^*(t) = 1 - \exp[-TFNR(t)/TI(t)] = 1 - e^{-AFMR(t)} \qquad (6.20)$$

By analogy to the ACF, PEM* provides a type of weighted average of the proportions ever marrying among the active cohorts.

### 6.8.2 Timing Effects on First Marriage in England and Wales and the United States

Schoen and Canudas-Romo (2005a) based their analyses on rates of first marriage derived from published vital statistics and census data. Cohort behavior was completed using the age-specific rates of the latest data year, 2001 for England and Wales and 2003 for the United States. For the cohort of 1972 (the latest or close to latest CPEM shown), observed data are available though age 29 for England and Wales and through age 31 for the United States. Since most first marriages in those countries are contracted before age 30, CPEM(1972) should be fairly accurate, even though partially estimated. A number of PEM* values are also dependent on estimated rates. Assume that a cohort TFNR can be accurately determined by rates up to age 35, and that few first marriages occur at ages 16 and below. Then PEM* values can be reliably found for England and Wales through 1983 (i.e. year 2001 minus 35 years for the oldest cohort plus 17 years for the youngest cohort) and for the United States through 1985.

Figure 6.4 shows the time trajectory of PEM* and period and cohort PEMs for women in England and Wales and the United States. Trajectories for men are qualitatively similar. The CPEM values for time t refer to the cohort born 29 years (about a generation) earlier. The overall pattern for both countries is a rise in proportions ever marrying up to around 1970 and a decline afterwards. Levels of first marriage are generally lower in England and Wales than in the United States, but England and Wales showed both a larger pre-1970 increase and a steeper post-1970 decline. For both populations, the cohort PEM fluctuates less than the period PEM, with the trajectory of the PEM* even smoother. Fluctuations in proportions ever marrying are associated with the two World Wars, the Depression of the 1930s, and the years immediately following World War II. Women in England and Wales in 1940, for example, have a PEM of 0.96 but a PEM* of 0.90. In the United States, PEM values notably understate average cohort behavior for women in the 1970s and 1980s. For example, in 1985 the U.S. PEM is 0.90 while the PEM* is 0.94.

Considerable interest has focused on recent declines in period marriage rates. The 1990–2000 decline for U.S. women was fairly small; the decline in PEM* from 0.93 to 0.91 was larger than the decline in PEM. For women in England and Wales, however, the fall in PEM values has been particularly large, going from 0.81 in 1990 to 0.69 in 2000. In contrast, the comparable PEM* values went from 0.85 to 0.77. Since those PEM* values are significantly influenced by estimated rates, two sets of additional calculations were done assuming that England and Wales first marriage rates either (i) decline linearly by 20% from 2001 to 2020 or (ii) increase linearly by 10% from 2001 to 2020. The resultant England and Wales PEM* values for 2001 show little change, varying only from 0.74 to 0.76. It is therefore quite likely that the year 2000 England and Wales PEM values for women are substantially influenced by timing effects, and that the average likelihood of first marriage in the cohorts active that year is about three-quarters rather than two-thirds.

## 6.9 TIMING EFFECTS ON DIVORCE

Divorce provides another instance where cohort tempo can affect period quantum. The classic study by Preston and McDonald (1979) found that the cohort probability a marriage would end in divorce increased smoothly (and exponentially) in the United States from 1867 to 1969, while period measures fluctuated markedly. Changes in divorce laws and practices provide

*Figure 6.4. Period and Cohort Proportion Ever Marrying (PEM) and Timing Adjusted* PEM* *Trajectories*

**A. Women in the England and Wales, 1910-2001**

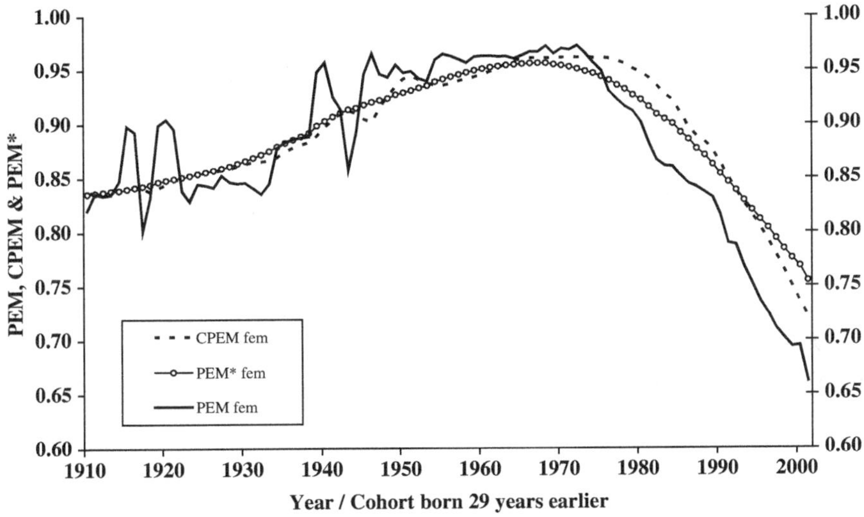


**B. Women in the United States, 1910-2003**

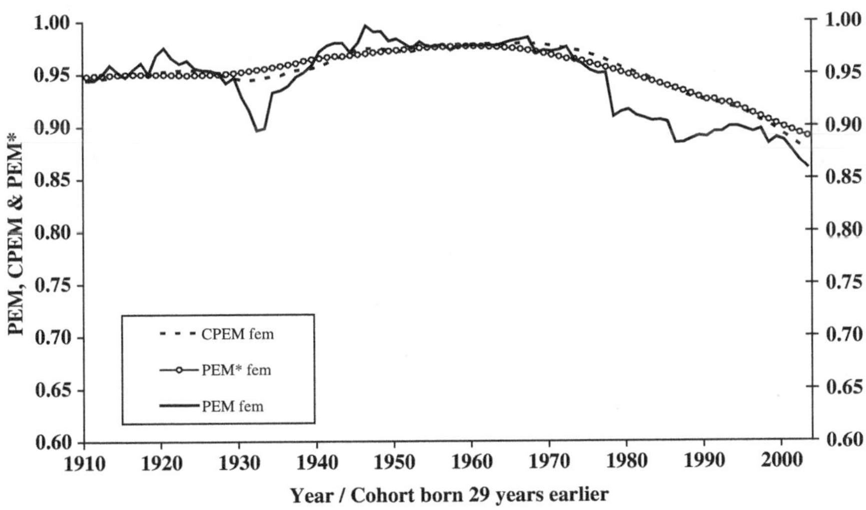


*Source:* Schoen and Jonsson (2005a, Figure 4). Reprinted with permission from *Mathematical Population Studies*, Taylor and Francis Group, LLC, http://www.tandf.co.uk/journals, © 2005.

another factor that can influence the timing of divorce (Schoen, Greenblatt and Mielke, 1975; Wright and Stetson 1978). In this section, we lean on the analysis of Schoen and Canudas-Romo (2006a), which examined divorce behavior in the twentieth century United States.

### 6.9.1 Measuring Timing Effects on Divorce

A number of studies (e.g. Schoen and Standish, 2001) have examined marriage and divorce behavior in the context of marital status life tables. Here we consider a multistate model with marital states Never Married (s), Presently Married (m), and Divorced (v). The focus is on ages 15 through 59, where mortality (and widowhood) can be ignored. In the model, the probability that a marriage ends in divorce (by age 60, but there is little divorce after that age) is simply the ratio of the total number of divorces to the total number of marriages. Let $d_{ij}(x)$ represent the number of life table cohort moves from state i to state j between the ages of x and $x+1$. Then the probability a marriage ends in divorce, PMED, can be written

$$\mathrm{PMED} = \Sigma_x\, d_{mv}(x)/\Sigma_x\, [d_{sm}(x) + d_{vm}(x)] \qquad (6.21)$$

where the sum ranges from age 15 through age 59.

To measure how the timing of cohort divorce affects the level of period divorce, we again follow the logic underlying the TI and ACF. In the cohort marital status life table for persons born in year $t-x$, let the number of divorces at age x (and time t) be denoted by $c_{mv}(x, t-x)$. The proportion $\beta$ can then be defined as

$$\beta(x, t) = c_{mv}(x, t-x)/\Sigma_y\, c_{mv}(y, t-x) \qquad (6.22)$$

Then the year t Timing Index is once more given by

$$\mathrm{TI}(t) = \Sigma_x\, \beta(x, t) \qquad (6.23)$$

Using the TI, the divorce rates observed in year t can be adjusted for timing effects. With an asterisk denoting an adjusted rate and $m_{mv}(x, t)$ representing the occurrence/exposure divorce rate for persons age x at time t, we have

$$m_{mv}{}^{*}(x, t) = m_{mv}(x, t)/\mathrm{TI}(t) \qquad (6.24)$$

The time t adjusted divorce rates are thus the observed divorce rates divided by TI(t). The time t tempo adjusted probability of divorce, PMED*(t), can be found from equation (6.21) using values from a period marital status life table for time t constructed using the $m_{mv}{}^{*}(x, t)$ as the age-specific rates of divorce.

### 6.9.2 Timing Effects on Divorce in the United States

Schoen and Canudas-Romo (2006a) examined twentieth century American divorce probabilities using published vital statistics and census data and previously constructed marital status life tables for the United States (cf. Schoen 1987; Schoen and Standish 2001; Schoen, Urton, Woodrow and Baj 1985). To complete cohort experience after 2000, the latest data year, the cohort PMED (CPMED) and the adjusted PMED* were calculated under three scenarios: (i) year 2000 age-specific divorce rates remain constant, (ii) year 2000 divorce rates decline linearly by 20% from 2000 to 2020 and then continue at that lower level, and (iii) year 2000 divorce rates increase linearly by 20% from 2000 to 2020 and then continue at that higher level.

Figure 6.5 shows the trajectories of PMED, PMED*, and CPMED for U.S. women over the years 1910 through 2000. The values for men are quite similar to those for women. For PMED* and CPMED, Figure 6.5 gives values

*Figure 6.5. Period and Cohort Probability of a Marriage Ending in Divorce (PMED) and Adjusted* PMED* *for Women Under Two Future Scenarios, Linear Increases and Decreases of 20% in Divorce Rates from 2000 to 2020 United States 1910–2000.*

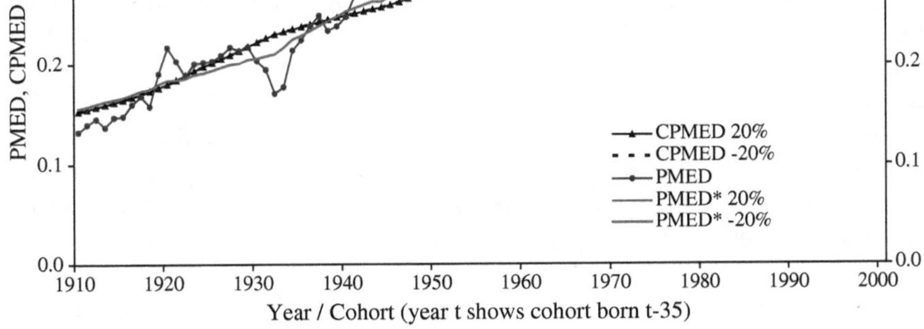


*Source:* Schoen and Canudas-Romo (2006a). Reprinted with permission from *Journal of Marriage and Family*, Blackwell © 2006.

under Scenarios (ii) and (iii), i.e. future decreases or increases of 20%. The CPMED value for time t refers to the cohort born 35 years earlier, as 35 years approximates the mean age at divorce.

Figure 6.5 indicates that the PMED has fluctuated markedly in the U.S. over the twentieth century, generally rising until about 1980 and then remaining roughly constant. The CPMED, lagged 35 years, and the PMED* have largely followed the same trend, though with much less irregularity. The fluctuations in PMED have led to sizeable timing effects on the probability of divorce. The 1946 PMED spike at the end of World War II has long been recognized as a period phenomenon, but tempo effects are evident in most decades. The timing of cohort divorce lowered the risk of divorce during the 1950s and 1960s, and raised it substantially from the early 1970s to the late 1980s. For example, in 1975 the PMED for women was 0.44 while PMED* was 0.38.

The PMED* peaked about 1990, and has not exhibited much of a trend since. It also demonstrated only a modest variability under the different scenarios presented. In 2000, the PMED* ranged from 0.44 to 0.46; the PMED was 0.42. Divorce rates imply that just under half of American marriages now end in a legal divorce, and the PMED* values provide no evidence that the risk of divorce has begun to fall.

## 6.10 SUMMARY

Timing effects are changes in the level of period measures caused by changes in the timing (but not the level) of cohort behavior. Both period and cohort perspectives are involved, because the idea of "postponing" or "advancing" events necessarily relates to cohort behavior across periods. Two ways to adjust period measures of fertility for timing effects are considered, the Bongaarts-Feeney and Average Cohort Fertility (ACF) approaches. The Bongaarts-Feeney method assumes that an observed period change in timing persists over a hypothetical cohort's reproductive years. The ACF method does not involve any strong assumptions, but uses lifetime data on the fertility of all actively reproducing cohorts to determine the extent to which cohort behavior is concentrated in a given year. The two measures are compared in the context of both models and the twentieth century experience of the United States. The results indicate that the ACF provides a sound and easily interpretable period measure that adjusts for timing effects, revealing the considerable impact those effects had on U.S. fertility during the Depression, the Baby Boom, and the 1970s Birth Dearth. The Bongaarts-Feeney adjusted TFR* is less satisfactory, frequently yielding values that substantially exaggerate or understate timing effects. The ACF approach, with modifications, can be applied to first marriage and divorce behavior. Doing so indicates that during the twentieth century timing effects substantially influenced period measures of divorce in the U.S. and first marriage behavior in both the U.S. and England and Wales.

A different tack toward dealing with timing issues is to make period rates specific to other important aspects of fertility, especially parity and duration since last birth. However, an analysis using age-parity-specific rates in a parity specific life table context indicates that such an approach can amplify rather than attenuate timing effects. In a fertility model where there are constant linear shifts over time in a fixed-shape age curve of fertility, timing effects appear to produce paradoxical circumstances. A population with every period NRR below 1 and every cohort NRR above 1 can arise, and represents a population where cohorts of decreasing size more than replace themselves.

## 6.11 EXERCISES

[See Appendix B for selected answers]

1. Ryder (1964) examined demographic "translation", i.e. how period TFRs can generate cohort TFRs and vice versa. His basic translation equation can be written

$$\mathrm{CTFR} = \mathrm{PTFR}/(1 - c)$$

   where c is the annual change in the <u>cohort</u> mean age at childbearing. Consider a Bongaarts-Feeney fertility model that ignores parity and has constant period linear shifts of r in its period fertility schedule (and thus in its period mean age at childbearing). Show that model implies $c = r$.
2. Consider a stable population where each age-specific net maternity rate is given by a continuous variant of equation (6.10), i.e.

$$\phi(x) = f(x)\, e^{sx}$$

   with $\int f(x)dx = 1$ and s an arbitrary parameter. Show that Lotka's $r = s$, and that the stable net maternity function $[e^{-rx}\phi(x)$ is constant for all s.
3. Consider a model where fertility has been constant over time and is uniform over all ages of childbearing, which range from $\alpha$ to $\beta$. Now assume that <u>cohort</u> fertility shifts upward by $\in$ years every year. Verify the Kim and Schoen (1999) relationship that, for periods after time $\beta$,

$$\mathrm{TFR} = \mathrm{CTFR}/(1 + \varepsilon)$$

4. Consider birth cohorts born during calendar years $t - 1$ and t. Assume that most deaths at age 0 in the year $t - 1$ birth cohort occur during the first

6 months of life, while most deaths at age 0 in the year t birth cohort occur during the second 6 months of life. Verify the argument in Horiuchi (2005) that such a timing pattern can produce a lower age 0 death rate during calendar year t than is experienced by either birth cohort.

5. Consider the dynamic linearly shifting fertility population of section 6.7, where the period NRR is less than one and the cohort NRR is greater than 1. Show logically that at any time t, the dynamic population is decreasing at a rate slower (i.e. less negative) than that of the stable population implied by the rates of time t.
6. Consider the linearly shifting fertility population examined in section 6.7 where the period NRR is always 1 and the cohort NRR is always greater than 1. Does such a population grow over time?

# CHAPTER 7

# DYNAMIC BIRTH-DEATH (MULTI-AGE) MODELS

## 7.1 INTRODUCTION

This chapter explores age-graded populations with rates of birth and death that change over time. The stable population, the principal model of mathematical demography, has the great strengths of logical closure, the ability to reflect the implications of any set of age-specific birth and death rates, and the profound insight that constant vital rates lead to both an exponential birth trajectory and an unchanging age composition. Yet the stable model is essentially static; it is based on unchanging vital rates and thus cannot reflect the dynamics of actual populations. Changing rates greatly complicate the relationship between vital rates and the birth trajectory as, in general, there is no way to simplify the product of a series of population projection matrices.

To seek regularities in models with changing rates, we first consider patterns in a dynamic model that does not explicitly include age. Assuming a constant generation length yields solutions for birth trajectories, including cyclical trajectories, from given sequences of net reproduction. Even a simple population with sinusoidal fluctuations in births can exhibit complex patterns of intergenerational transfers. Next we examine two models, the "metastable" and the "intrinsically dynamic", that generalize the stable model, recognize age, and yield solutions for birth trajectories from changing rates. Finally, two methods for determining age-specific rates consistent with a given birth trajectory are examined.

## 7.2 MODELS WHERE THE AGE-SPECIFIC RATES ARE IMPLICIT

Pioneering analytical work on populations with changing vital rates was done by Coale (1972) and Lee (1974). Kim (1987) found a general algebraic solution relating rates to births, but in most instances it is too complex to render in closed form. Cyclically stable models, produced by a fixed sequence of rates that repeats indefinitely, were introduced by Skellam (1967) and examined by Namboodiri (1969), Coale (1972), Tuljapurkar (1985; 1990), and Schoen and

Kim (1994). Nonlinear models, where the rates are influenced by population size and composition, have begun to be explored (Tuljapurkar 1990; Frauenthal and Swick 1983; Wachter and Lee 1989; Wachter 1991), but such analyses frequently lead to great complexity if not "chaos", and are largely beyond the scope of this chapter (cf. May 1974; Strogatz 1994; Alligood, Sauer and Yorke 1996).

### 7.2.1 Considering Models With a Constant Generation Length

Schoen and Kim (1997) developed a technique for finding the birth trajectory that results from any given sequence of net reproduction. This and the following subsection draw heavily on that work.

Let g(t) be the number of births between times t and t + dt, scaled so that g(0) = 1. Denote the net maternity rate between the ages of x and x + dx at time t by $\phi(x, t)$, where

$$\phi(x, t) = f(x, t)\ p(x, t) \tag{7.1}$$

with f(x, t) the fertility rate at age x and time t and p(x, t) the probability of surviving to attain age x at time t. General population renewal equation (1.27) can then be rewritten

$$1 = \int [g(t - x)/g(t)]\phi(x, t)\ dx \tag{7.2}$$

and the Net Reproduction Rate at time t, R(t), is given by

$$R(t) = \int \phi(x, t)\ dx \tag{7.3}$$

The Mean Value Theorem can be applied to equation (7.2), and using equation (7.3) we have

$$R(t) = g(t)/g(t - A(t)) \tag{7.4}$$

where A(t) is the applicable mean value at time t. Schoen and Kim (1997) assumed that net maternity changed in such a way that A(t) remained constant over time, yielding

$$R(t) = g(t)/g(t - A) \tag{7.5}$$

Recalling stable population equation (1.38), A can be interpreted as a generation length.

Looking forward over time, with $g(0) = 1$, equation (7.5) leads to the relationship

$$g(nA) = R(A)\ R(2A) \ldots R(nA) \tag{7.6}$$

where n can be any positive integer. Equation (7.6) can be rewritten as an exponentiated sum of logs, i.e. as

$$g(nA) = \exp[\Sigma \ln R(jA)] \tag{7.7}$$

where the sum over j ranges from 1 to n. Equation (7.7) expresses a birth sequence in terms of NRRs, and letting $t = nA$, we can use it to explore times of any length.

Consider the case where R(t) is constant over time at value R. Then

$$g(t) = \exp[(\ln R)\ t/A] \tag{7.8}$$

indicating stable (exponential) population growth at rate $r = (\ln R)/A$. From equation (1.38), it is evident that in this case A is Lotka's mean generation length Ŧ.

Let us now consider the case where R(t) increases exponentially, i.e. where $R(t) = e^{ht}$. The sum in equation then leads to

$$g(t) = \exp[(ht + ht^2)/(2A)] \tag{7.9}$$

An exponential increase in R(t) thus implies a quadratic exponential increase in births. In general, equation (7.7) implies that if the NRR increases by an exponentiated mth degree polynomial, then g(t) increases according to an exponentiated $(m + 1)$st degree polynomial.

Two caveats apply to the general solution in equation (7.7). First, because that equation is based on discrete time units, it is necessary to verify that the result holds at any time t. That can be done using equation (7.5), and in the cases presented here (and in Schoen and Kim 1997) the ratio $g(t)/g(t - A)$ gives the original R(t) at all t. Second, although the Mean Value Theorem guarantees that at least one real, non-negative schedule of age-specific net maternity rates exists at every time, there may be only one such schedule. As time approaches A years after an extremum of g(t), the variance of $\phi(x, t)$ declines markedly. Exactly A years after an extremum, all net maternity is concentrated at age A. Even that worst case, however, involves nothing more than assuming that all net maternity is concentrated at the mean.

### 7.2.2 Cyclical Models With a Constant Generation Length

Let us explore the case where the NRR is cyclical, specifically where

$$R(t) = \exp[b\ \sin(\omega t)] \qquad (7.10)$$

with $\omega$ representing the cycle frequency. By definition, the cycle length, T, satisfies the relationship $\omega = 2\pi/T$. Schoen and Kim (1997), using an analytical summation of equation (7.7) and the cotangent identity $\cot(x/2) = \sin(x)/[1-\cos(x)]$, found that

$$g(t) = g(0)\ \exp[(b/2)\{\sin(\omega t) + [1-\cos(\omega t)]\cot(\omega A/2)\}] \qquad (7.11)$$

Equation (7.11) provides an exact birth trajectory for the cyclical NRR function in equation (7.10), and Figure 7.1 shows birth and NRR trajectories for two combinations of parameter values.

Qualitatively, relationships in the cyclical model depend largely on the ratio of cycle length T to generation length A. The focus is on models where $T > A/2$, i.e. the cycle length is about 15 years or more. That includes most models of demographic interest, and there is only one case (i.e. $T = A$) where g(t) is not defined. If $\omega A/2 = n\pi$, n a non-negative integer, $\cot(\omega A/2)$

*Figure 7.1. Exponentiated Sinusoidal NRR Trajectory, R(t), and Associated Birth Trajectory, g(t), for Generation Length A = 30, Amplitude b = 0.2, and Selected Cycle Lengths, T*

**FIGURE 1a T = 2A = 60** **FIGURE 1b T = 4A = 120**

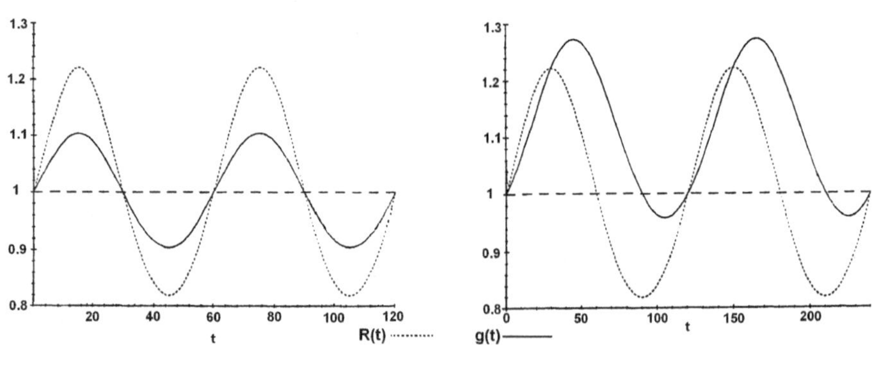


*Source:* Schoen and Kim (1997, Figure 1cd). Reprinted with permission from *Mathematical Population Studies*, Taylor and Francis Group, LLC, http://taylorandfrancis.com © 1997.

becomes infinite and g(t) is undefined. In contrast, if $\omega A/2 = (n + 1/2)\pi$, then $\cot(\omega A/2) = 0$ and the birth trajectory reduces to

$$g(t) = [R(t)]^{1/2} \tag{7.12}$$

If equation (7.12) applies, there is no phase difference between g(t) and R(t) and both are exponentiated sinusoids (see Figure 7.1a). For other generation lengths, the extrema of g(t) can be found by setting the time derivative of g(t) equal to zero. If $t_{gmax}$ identifies the time when g(t) attains its first maximum after $t = 0$,

$$\omega\, t_{gmax} = \begin{cases} \pi - \omega A/2 & \text{if } A < T \\ 2\pi - \omega A/2 & \text{if } A/2 < T < A \end{cases} \tag{7.13}$$

Defining phase difference $\theta$ as the difference in time between the maximum of g(t) and the maximum of R(t) expressed as a fraction of cycle length, we have

$$\theta = (t_{gmax} - t_{Rmax})/T \tag{7.14}$$

where $t_{Rmax}$ is the time where R(t) attains its first maximum after $t = 0$, i.e. T/4. When $\theta > 0$, g(t) leads R(t), and in general

$$\theta = \begin{cases} 1/4 - A/(2T) & \text{if } A < T \\ 3/4 - A/(2T) & \text{if } A/2 < T < A \end{cases} \tag{7.15}$$

The size of $\theta$ is bounded by $\pm\, 1/4$. For $A < T < 2A$, $\theta$ is negative [and R(t) leads g(t)], and for $T > 2A$, $\theta$ is always positive (e.g. Figure 7.1b). Since the maxima and minima of both g(t) and R(t) differ by exactly T/2, $\theta$ is the phase difference between both their maxima and minima.

Relative amplification also depends on the ratio of T to A, but is influenced by amplitude parameter b as well. With $b > 0$, the maximum and minimum values of R, $R_{max}$ and $R_{min}$, are given by

$$R_{max} = e^{b};\quad R_{min} = e^{-b} \tag{7.16}$$

The maximum and minimum values of birth trajectory g, $g_{max}$ and $g_{min}$, are given by

$$g_{max} = \exp[b\ \cot(\omega A/4)/2];\quad g_{min} = \exp[-b\ \tan(\omega A/4)/2] \tag{7.17}$$

for $T > A$, and by those same relationships with minimum and maximum switched for $A/2 < T < A$. If the index of relative amplification, M, is the ratio of $g_{max}$ to $g_{min}$ divided by the ratio of $R_{max}$ to $R_{min}$, then

$$M = \begin{cases} \exp[b\{\csc(\omega A/2) - 2\}] & \text{if } A < T \\ \exp[-b\{\csc(\omega A/2) + 2\}] & \text{if } A/2 < T < A \end{cases} \tag{7.18}$$

or in terms of the phase difference

$$M = \exp[b\{\sec(2\pi\theta) - 2\}] \tag{7.19}$$

Equation (7.19) applies at all cycle lengths, and indicates that M varies directly with the absolute value of $\theta$. When the phase difference is zero, M is at its minimum value, $e^{-b}$. M has no maximum value, and goes to $\infty$ as T approaches $A/n$ (n a nonnegative integer). There is no relative amplification (i.e. $M = 1$) when $\theta$ is $\pm(1/6)$ radians. For $T > A/2$, that occurs when T is 6A, 6A/5, 6A/7, and 6A/11. The ratio of NRRs exceeds the ratio of births (i.e. $M < 1$) over a substantial span of demographically likely cycle lengths, including 1.2A to 6A.

Schoen and Kim (1997), following on work by Wachter (1991; 1994), noted a connection between the exponentiated sinusoidal model and the Easterlin Hypothesis. Easterlin (1980) argued that large birth cohorts encounter economic disadvantages because of their size, leading them to have lower fertility, while small cohorts are economically advantaged and have higher fertility. In the model of equations (7.10) and (7.11), assume that a cohort's fertility is concentrated at mean age A. The NRR at time t then reflects the reproductive level of the cohort born at time t-A. When there is no phase difference between R(t) and g(t), for example when T is 2A or 2A/3, behavior in the exponentiated sinusoidal model follows the Easterlin Hypothesis. As illustrated in Figure 1a, where $T = 2A$, large cohorts have the fewest children and small cohorts have the most children.

### 7.2.3 Intergenerational Relationships in a Simple Model With Sinusoidal Births

A stationary population has often been used as a long term, zero growth model, but observers as far back as Malthus believed that populations would oscillate about their long term equilibrium (cited in Lee 1997, p1092). Economic-demographic interactions can give rise to a number of cyclical patterns, including Easterlin's cycles of two generations (cf. Lee 1997). Epidemics and climate cycles can also contribute. At the individual level, fluctuations in the number of births can have substantial impacts. Preston (1984) saw a shift in transfer payments from the young to the old, Menken (1985) spoke of a dependency bind where middle aged persons faced simultaneous demands from children and aging parents, and a number of observers (e.g. Keyfitz 1988) saw a rise in intercohort inequities in pension benefits.

To illustrate some of the intergenerational dynamics generated by birth fluctuations, we follow Schoen and Jonsson (2006) and consider a simple, basic model. Assume a cyclically stationary population, where there is no long term growth and the vital rates repeat every T years. Fertility rates are such as to yield birth cohorts whose size in year t, g(t), is given by

$$g(t) = 1 + b\ \sin(\omega t) \tag{7.20}$$

where $-1 < b < 1$ represents the amplitude of the sine wave, $\omega$ is its frequency, $T = 2\pi/\omega$, and the model is scaled so that $g(0) = 1$. Every person born survives to exact age 90, and then dies. The total population size at time t, P(t), is then

$$P(t) = 90 + [b/\omega]\{\cos(\omega[t-90]) - \cos(\omega t)\} \tag{7.21}$$

It follows that the size of the population is bounded by $90 \pm 2b/\omega$ [or by $90 \pm bT/\pi$].

To distinguish between the economically active and inactive (dependent) components of the population, we arbitrarily set the active ages as 20 to 65, half the lifespan. No distinction is made between young age dependents (at ages 0 to 20) and old age dependents (at ages 65 and over). The total population dependency ratio at time t, D(t), is then

$$D(t) = [P(0, 20, t) + P(65, 25, t)]/P(20, 45, t) \tag{7.22}$$

where P(x, n, t) represents the population between the ages of x and $x + n$ at time t.

The analysis in Schoen and Jonsson (2006) indicates that qualitatively, the dynamics of the model are determined by cycle length T. To focus on the range of greatest demographic interest, we restrict T to between 15 and 105 years. Amplitude b increases (or decreases) the magnitude of the effects in close to a linear fashion. Accordingly, b is set equal to 0.3, on the argument that it is plausible to have the NRR vary from 0.7 to 1.3.

Figure 7.2 shows how the maximum dependency ratio varies with T. (The minimum dependency ratio is approximately the reciprocal of the maximum.) The maximum value of D is the base value of 1 whenever T is an integral divisor of 45, that is at graphed values of 15, 22.5, and 45 years. For $15 < T < 22.5$, the maximum D is less that 1.1, while for $22.5 < T < 45$ it reaches 1.195 when $T = 35$. The maximum D increases steadily for $T > 45$, reaching a peak of 1.478 when $T = 83$. For larger values of T, the maximum

*Figure 7.2. Maximum Dependency Ratio (D) When Amplitude (b) is 0.3, for Values of Cycle Length (T) from 15 to 105 Years in a Cyclical Stationary Population Model*

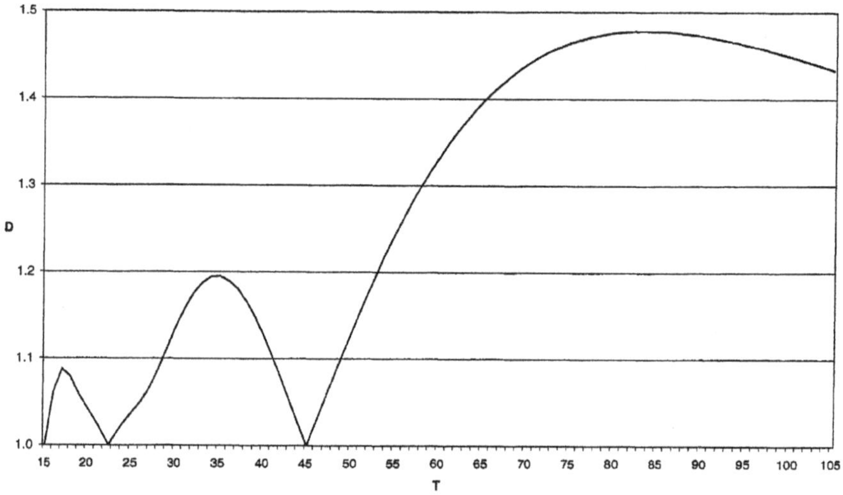


*Source:* Originally published as Figure 1 in Robert Schoen and Stefan Jonsson (2006). "Some Intergenerational Transfer Implications of Birth Fluctuations" in "*Allocating Public and Private Resources across Generations*", Anne H. Gauthier, C.Y. Cyrus Chu and Shripad Tuljapurkar, eds. pp. 279–290. Reprinted with kind permission from Springer Science and Business Media.

D declines slowly, eventually approaching 1. Within each T year cycle, the dependency burden can very considerably. For T = 83, D goes from a high of 1.478 to a low of 0.664, indicating that in some years the dependency burden is more than twice that of other years.

A major concern is the effect of cyclical fluctuations on intergenerational transfer systems, especially pension funds. Such transfers can be examined in the simple, cyclically stationary model of equation (7.20). Schoen and Jonsson (2006) considered two common types of transfer systems. The first, known as defined contribution plans, specify the contributions to be made. Here we assume that each active person contributes $1 each year, and refer to the plan as "Pay $45". The second type, known as defined benefit plans, specify the benefit to be received. Here we assume that each dependent receives $1 a year, and refer to the plan as "Get $45". As is more or less the case with national pension systems, the program functions on a "pay as you go" basis. Each year the system distributes the funds that are paid in. Thus in the Pay $45 plan, each dependent gets $(1/D(t)) in year t, while in the Get $45 plan each economically active person contributes $D(t) in year t.

Table 7.1 shows maximum dollar gains and losses under Pay \$45 and Get \$45 plans by cycle length. At cycle lengths characterized by large maximum dependency ratios, the year in which one is born in the cycle matters greatly. In Pay \$45 plans, large cohorts generally get less than they contribute while small cohorts get more. In Get \$45 plans, the reverse is true, and members of large cohorts get more than they pay in over their lifetimes. When $T = 35$, the most favored birth cohort in a Pay \$45 plan (born in year 26 of the cycle) receives \$49.00, while the least favored cohorts (born in years 8 and 9 of the cycle) receive only \$41.91, less than 86% of that amount. When $T > 55$, the discrepancies are generally even larger. Even in this simple and predictable model, both defined benefit and defined contribution plans lead to substantial inequities between cohorts. Weaving them together to produce a "fair" plan for an actual population is clearly a daunting task (cf. De Santis 2003).

*Table 7.1. Maximum Dollar Gain and Loss Values (Benefits Minus Contributions) Under Pay \$45 and Get \$45 Plans, When Amplitude (b) is 0.3, By Cycle Length (T) in a Cyclically Stationary Population*

| T | Pay \$45 | | | Get \$45 | | |
|---|---|---|---|---|---|---|
| | Maximum Gain | Maximum Loss | Gain − Loss | Maximum Gain | Maximum Loss | Gain − Loss |
| 20 | 0.21 | −0.17 | 0.38 | 0.17 | −0.21 | 0.38 |
| 25 | 0.18 | −0.16 | 0.34 | 0.16 | −0.19 | 0.34 |
| 30 | 1.40 | −1.03 | 2.43 | 1.03 | −1.40 | 2.43 |
| 35 | 4.00 | −3.09 | 7.09 | 1.34 | −1.87 | 3.21 |
| 40 | 1.94 | −1.42 | 3.36 | 0.50 | −0.69 | 1.19 |
| 45 | 0 | 0 | 0 | 0 | 0 | 0 |
| 50 | 2.05 | −1.55 | 3.60 | 0.54 | −0.70 | 1.24 |
| 55 | 6.82 | −5.01 | 11.83 | 1.92 | −2.45 | 4.37 |
| 60 | 12.05 | −8.36 | 20.41 | 3.66 | −4.63 | 8.29 |
| 65 | 16.04 | −10.62 | 26.66 | 5.37 | −6.79 | 12.16 |
| 70 | 18.12 | −11.73 | 29.85 | 6.83 | −8.70 | 15.53 |
| 75 | 18.30 | −11.93 | 30.27 | 7.95 | −10.26 | 18.21 |
| 80 | 17.09 | −11.50 | 28.59 | 8.74 | −11.45 | 20.19 |
| 85 | 15.14 | −10.64 | 25.78 | 9.25 | −12.32 | 21.57 |
| 90 | 12.90 | −9.52 | 22.42 | 9.52 | −12.90 | 22.42 |
| 95 | 10.69 | −8.28 | 18.97 | 9.61 | −13.25 | 22.86 |
| 100 | 8.66 | −6.99 | 15.65 | 9.57 | −13.40 | 22.97 |
| 105 | 6.88 | −5.72 | 12.60 | 9.42 | −13.42 | 22.84 |

*Source*: Originally published as Table 1 in Robert Schoen and Stefan Jonsson (2006). "Some Intergenerational Transfer Implications of Birth Fluctuations" in "*Allocating Public and Private Resources across Generations*", Anne H. Gauthier, C.Y. Cyrus Chu and Shripad Tuljapurkar, eds. pp. 279–290. Reprinted with kind permission from Springer Science and Business Media.

## 7.3 METASTABILITY, A CLOSED FORM EXTENSION OF STABILITY

The constant generation models of section 7.2 do not explicitly include age, an essential aspect of conventional demographic models. In particular, while equation (7.9) indicates that exponentially increasing net maternity leads to an exponentiated quadratic birth sequence, it does not provide a set of underlying age-time specific net maternity schedules. Doing so is not a simple matter. Coale (1972) examined the problem, but was able to find only an approximate solution. Feichtinger and Vogelsang (1978) and Vogelsang and Feichtinger (1979) extended the analysis, but could not find a closed form relationship between the exponentially increasing rates and the quadratic exponential birth sequence.

In this section, we draw on Schoen and Kim (1994a), Kim and Schoen (1996), and especially Schoen and Jonsson (2003), which provide such a closed form relationship and examine some of its implications. That last paper called the model Quadratic Hyperstable (QH), but here it is termed "metastable" to set it apart and emphasize that it goes beyond classical stability to incorporate the implications of exponentially changing vital rates through a set of closed form relationships that include the stable model as a special case.

### 7.3.1 Deriving the Discrete Metastable Model

The discrete formulation provides the easiest access to the metastable model. Consider an $n \times n$ time-varying population projection (Leslie) matrix of the form

$$\mathbf{A}_t = \begin{bmatrix} f_1 c\, k^t & f_2 c^2 k^{2t} & f_3 c^3 k^{3t} & \dots & f_n c^n k^{nt} \\ 1 & 0 & 0 & \dots & 0 \\ 0 & 1 & 0 & \dots & 0 \\ 0 & 0 & \dots & 1 & 0 \end{bmatrix} \tag{7.23}$$

The $f_j$ factors provide the base age pattern of net maternity, with $\Sigma_j\ f_j = 1$. Parameter c is the dominant eigenvalue of the matrix at time 0, and gives the initial (implicit) stable growth level. The survivorship values in the subdiagonal are set to 1. Doing so greatly simplifies the algebra at very little cost, though it must be borne in mind that the resultant sequence of Leslie matrices generates a birth trajectory rather that a population trajectory, leaving mortality to be added later.

The powers of k in the first row show how net maternity varies exponentially over both age and time. When $k = 1$, the $\mathbf{A}_t$ generate a stable population with growth rate c. When net maternity increases over time, i.e. $k > 1$, values at the higher ages increase more, a reasonable pattern because higher fertility implies more higher order births, which arise largely at older ages. The opposite holds when net maternity decreases over time, i.e. when $k < 1$, a pattern that is consistent with observed declines in fertility. Analytically, that exponential pattern of change in fertility has been applied by a number of researchers (including Sykes 1973; Mitra 1976; and Schoen and Kim 1996), was incorporated in the structure of the Coale-Trussell Model Fertility Schedules (Coale and Trussell 1974), and was employed in equation (6.10) to systematically vary the pattern of fertility. That pattern of net maternity change has two important implications. First, it produces a constant proportional distribution of births by age of mother even as the level of net maternity changes over time. Second, net maternity always increases (or decreases) at a constant pace over time.

Because of the form of the $\mathbf{A}_t$ in equation (7.23), Sykes (1973) noted that

$$\mathbf{A}_t = c\, k^t\, \mathbf{U}_t\, \mathbf{F}\, \mathbf{U}_t^{-1} \tag{7.24}$$

where $\mathbf{U}_t$ is an $n \times n$ diagonal matrix whose jth diagonal element is $(c\, k^t)^{-(j-1)}$, the jth element of the dominant right eigenvector of $\mathbf{A}_t$. The $\mathbf{F}$ matrix is of the form

$$\mathbf{F} = \begin{bmatrix} f_1 & f_2 & f_3 & \dots & f_n \\ 1 & 0 & 0 & \dots & 0 \\ 0 & 1 & 0 & \dots & 0 \\ 0 & 0 & \dots & 1 & 0 \end{bmatrix} \tag{7.25}$$

As it is a row stochastic population projection matrix (i.e. all of its rows sum to 1), its dominant eigenvalue ($\lambda_F$) equals 1. The "Sykes transformation" of equation (7.24) rewrites $\mathbf{A}_t$ in terms of its dominant eigenvalue ($c\, k^t$), a time varying diagonal matrix ($\mathbf{U}_t$) and its inverse, and constant matrix $\mathbf{F}$.

Let us examine the product matrix that projects the initial (time 0) population to time t, that is

$$\mathbf{M}_{0,t} = \mathbf{A}_t\, \mathbf{A}_{t-1}\, \mathbf{A}_{t-2} \dots \mathbf{A}_1 \tag{7.26}$$

Crucially, within this sequence of matrix multiplications, the matrix product $\mathbf{U}_{\tau+1}^{-1}\, \mathbf{U}_\tau$ always reduces to a constant diagonal matrix, $\mathbf{K}$, whose jth diagonal element is $\mathbf{k}^{j-1}$. Using equation (7.24), equation (7.26) can be rewritten as

$$\mathbf{M}_{0,t} = c^t\, k^{t(t+1)/2}\, \mathbf{U}_t\, [\mathbf{FK}]^t\, \mathbf{U}_0^{-1} \tag{7.27}$$

Equation (7.27) is the key to the metastable model, as it expresses the product of the $\mathbf{A}_t$ in terms of three factors: (1) the product of scalar growth rates c and k from times 1 through t; (2) diagonal matrices reflecting the stable age compositions of the initial (time 0) and final (time t) Leslie matrices; and (3) a constant Leslie-form matrix raised to the tth power. The time invariant matrix **S** that synthesizes **F** and **K** can be written

$$\mathbf{S} = \mathbf{F}\,\mathbf{K} = \begin{bmatrix} f_1 & f_2 k & f_3 k^2 & \dots & f_n k^{n-1} \\ 1 & 0 & 0 & \dots & 0 \\ 0 & k & 0 & \dots & 0 \\ 0 & 0 & \dots & k^{n-2} & 0 \end{bmatrix} \qquad (7.28)$$

The emergence of constant matrix **S** makes convergence in the metastable model equivalent to stable convergence. After a sufficiently long time, say $\tau$ intervals, equation (2.3) implies

$$\mathbf{S}^{\tau} = \lambda_s^{\tau}\, \mathbf{u_s}\, \mathbf{v'_s} \qquad (7.29)$$

where $\lambda_s$ is the dominant eigenvalue of **S**, $\mathbf{u_s}$ is the dominant right eigenvector of **S**, and $\mathbf{v'_s}$ is the dominant left eigenvector of **S**. From the structure of **S**, $\lambda_s$ is greater than, equal to, or less than 1 as k is greater than, equal to, or less than 1. (The eigenstructure of **S** is further explored in Exercises 2 and 3.)

When equation (7.29) applies, the product matrix becomes

$$\mathbf{M_{0,t}} = c^t\, k^{t(t+1)/2}\, \lambda_s^{t}\, \mathbf{U_t}\, \mathbf{u_s}\, \mathbf{v'_s}\, \mathbf{U_0}^{-1} \qquad (7.30)$$

which captures the long term dynamics of the metastable model. From the projection relationship $\mathbf{x_t} = \mathbf{M_{0,t}}\, \mathbf{x_0}$, we have

$$\mathbf{x_t} = c^t\, k^{t(t+1)/2}\, \lambda_s^{t}\, \mathbf{U_t}\, \mathbf{u_s}\, (\mathbf{v'_s}\, \mathbf{U_0}^{-1}\, \mathbf{x_0}) \qquad (7.31)$$

which provides the long term size and composition of the metastable population.

The product $Q_M = (\mathbf{v'_s}\, \mathbf{U_0}^{-1}\, \mathbf{x_0})$ is a scalar analogous to the product $Q = (\mathbf{v'x_0})$ in equation (2.4), and represents the contribution of initial population $\mathbf{x_0}$ to the size of the time t metastable population. In convergence to metastability, as in convergence to stability, the long term effect of the size and composition of the initial population is captured by a single scalar constant.

The age composition of the metastable population at time t (more precisely the size of the birth cohorts of the recent past relative to the number of births at time t) is given by the product $\mathbf{U_t}\, \mathbf{u_s}$. At time t, that n-age group vector can be written

$$\mathbf{U_t}\, \mathbf{u_s} = \begin{bmatrix} 1 \\ (c\lambda_s)^{-1}\, k^{-t} \\ (c\lambda_s)^{-2}\, k^{-2t}\, k \\ \vdots \\ (c\lambda_s)^{-(n-1)}\, k^{-t(n-1)}\, k^{(n-1)(n-2)/2} \end{bmatrix} \qquad (7.32)$$

The metastable age composition combines the time-varying composition associated with $\mathbf{A_t}$ (i.e. $\mathbf{U_t}$) and the fixed age composition associated with $\mathbf{S}$ (i.e. $\mathbf{u_s}$). As the jth age group includes the factor $k^{-t(j-1)}$, the relative size of the older age groups increases over time when $k < 1$. The jth age group also has a factor of k to the power $j^2$. When $k > 1$, that factor increases the size of the jth age group relative to the first age group, reflecting slower population growth in the past than at time t. The reverse is the case when $k < 1$.

The time t metastable population grows at rate $(c\lambda_s\ k^t)$. That growth rate increases over time when $k > 1$, and differs from the long term growth rate implied by $\mathbf{A_t}$ by a factor of $\lambda_s$. If fertility is declining, i.e. both k and $\lambda_s$ are less than 1, the time t metastable population has a slower pace of growth than the stable population with the same vital rates. From equation (7.30), the birth trajectory (i.e. the number in the first age group) of the metastable population is given by

$$g_t = g_0(c\lambda_s)^t k^{t(t+1)/2} \tag{7.33}$$

The long term birth trajectory is an exponentiated quadratic based on parameter k accompanied by a linear exponential based on $(c\lambda_s)$.

Growth factor $\lambda_s$, which emerges in the derivation, further augments the pace of growth in the number of births when $k > 1$ and retards it when $k < 1$. That growth factor essentially reflects the different histories of the metastable and associated stable populations. When $k > 1$, the size of the cohorts of reproductive age (especially those at the higher reproductive ages where the fertility rates increase most rapidly) relative to the size of the cohort age 0 is generally larger in the metastable than in the associated stable population, because the metastable population grew more slowly in the past. The reverse is true when $k < 1$. That yields different population proportions exposed to the same regime of vital rates, and thus different rates of growth.

The above closed form equations fully specify a discrete, rate generated, dynamic model in which fertility can increase or decrease by any fixed rate. The underlying vital rates and the size and structure of the metastable population at any time are provided through exact relationships. As we see below, the model can be used to analyze fertility transitions in general, as well as gradual transitions to stationarity and the population growth associated with them.

### 7.3.2 The Continuous Metastable Model

The discrete equations of the previous subsection have their continuous counterparts. With net maternity rates given by the $\phi(x,t)$ of equation (7.1) and known initial rates $\phi(x, 0)$, let the pattern of net maternity change over age and time be given by

$$\phi(x, t) = \phi(x, 0)\ e^{hxt} = \phi_0(x)\ e^{rx}\ e^{hxt} \tag{7.34}$$

where h is a known constant and $\phi(x, 0) = \phi_0(x)\ e^{rx}$. Note that net maternity changes with the exponentiated product of both age and time. Since the time 0 stable population characteristic equation is

$$1 = \int e^{-rx}\ \phi(x, 0)\ dx \tag{7.35}$$

we must have $1 = \int \phi_0(x)\ dx$ and r as the time 0 stable population intrinsic growth rate.

The number of births at time t in the metastable population is then given by

$$g(t) = g(0)\ \exp[(r+s)t + ht^2/2] \tag{7.36}$$

which satisfies $g(t) = \int g(t-x)\ \phi(x, t)dx$ when parameter s is defined by

$$1 = \int \exp[-sx + hx^2/2]\ \phi_0(x)\ dx \tag{7.37}$$

The quantity integrated in equation (7.37) represents the metastable fixed fraction of births to mothers age x. Continuous equation (7.37) is analogous to the discrete characteristic equation of matrix **S**, i.e.

$$1 = \Sigma_j\ f_j\ k^{j(j-1)/2}\ \lambda_s^{-j} \tag{7.38}$$

where $f_j$ corresponds to $\phi_0(x)$, $\lambda_s$ to exp[s], and $k^{j(j-1)/2}$ to $\exp[hx^2/2]$. Equations (7.34) through (7.37) fully specify the continuous metastable model, and reduce to the stable model when change parameter $h = 0$.

Let us explore some relationships in the continuous model. With birth function g(t) and p(x, t) as the probability of surviving to age x at time t, the number of persons age x at time t, relative to g(t), can be represented by n(x, t) and written as

$$n(x, t) = \exp[-x(r+s+ht) + hx^2/2]\ p(x, t) \tag{7.39}$$

Equation (7.39) is analogous to discrete equation (7.32).

Age-specific growth rate r(x, t) is defined in equation (4.18). If we assume constant mortality, it can be found by differentiating equation (7.36) and evaluating the derivative at $(t-x)$. Kim and Schoen (1996) found the result

$$r(x, t) = r + s + h(t - x) \tag{7.40}$$

Thus when $h > 0$, the age-time-specific growth rate increases over time at all ages, though at any given time it decreases over age.

Consider the time t associated stable population. Its intrinsic growth rate at time t, $r_A(t)$, is defined, using equation (7.34), by characteristic equation

$$1 = \int \exp[-x\{r_A(t) - ht\}]\ \phi(x, 0)\ dx \tag{7.41}$$

Since $r_A(0)$ is r, equation (7.41) implies that

$$r_A(t) = r + ht \tag{7.42}$$

or that the intrinsic growth rate of the associated stable population increases linearly over time. That relationship also implies that the proportional distribution of births by age of mother is the same in every associated stable population, as the changes in $r_A(t)$ and $\phi(x, t)$ offset one another. That fixed proportional distribution of births, however, differs from its metastable counterpart.

Age-specific momentum at time t, $\Omega(x, t)$, is defined in equation (4.49) as the ratio of S(x, t), the number of persons in the stable equivalent population at age x and time t, divided by N(x, t), the number of persons in the population at age x and time t. Following Kim and Schoen (1996), for a metastable population at time t, S(x, t) can be written

$$S(x, t) = Q(t)\ \exp[-x(r + ht)]\ p(x) \tag{7.43}$$

where p(x) reflects survivorship to age x under the assumed constant mortality rates. Q(t), the time t stable equivalent number of births, is given by

$$Q(t) = \int N(x, t)\ V(x, t)\ dx \tag{7.44}$$

The age-specific reproductive value at time t, V(x, t), is provided by

$$V(x, t) = \frac{\int_x^\infty e^{-a(r+ht)}\ \phi(a, t)\ da}{e^{-x(r+ht)}\ p(x)\ \int_0^\infty \int_x^\infty e^{-a(r+ht)}\ \phi(a, t)\ da\ dx} \tag{7.45}$$

Using equation (7.34) and the metastable fixed proportional distribution of births by age of mother, equation (7.45) can be written

$$V(x, t) = V(x, 0)\ e^{hxt} \tag{7.46}$$

which indicates that V(x, t) changes in the same way that $\phi(x, t)$ changes. The double integral in the denominator of equation (7.45), which represents the mean age of stable net maternity, is constant over time. The fraction of the total time t reproductive value arising at age x, $Q(x,t) = N(x,t)V(x,t)$ is, using equation (7.46)

$$Q(x,t) = Q(x,0)\ \exp[(r+s)t + ht^2/2] \qquad (7.47)$$

Integrating over age we have

$$Q(t) = Q(0)\ \exp[(r+s)t + ht^2/2] \qquad (7.48)$$

which indicates that Q(t) increases in the same manner as g(t) does. Using equation (7.36) and equations (7.43) through (7.48), we find the Kim and Schoen (1996) result

$$\Omega(x,t) = [Q(0)/g(0)]\ \exp[sx - hx^2/2] = S(x,0)/N(x,0) \qquad (7.49)$$

Age-specific momentum is time invariant in metastable models. At every age, the ratio of the associated stable population to the metastable population remains constant over time.

The Kullback distance, K(t), defined in equation (2.11), measures the distance between a given population and the stable population implied by the prevailing vital rates. K(t) is a function of $\Omega(x,t)$ and $q(x,t) = Q(x,t)/Q(t)$, neither of which changes over time in the metastable case. The Kullback distance of a metastable population is thus constant over time. In the long term, the metastable population never moves closer to nor further from its associated stable population, even though the structure and vital rates of both populations are continually changing.

### 7.3.3 Modeling Fertility Transitions and Their Associated Momentum

The metastable model facilitates the analysis of transitions from one set of vital rates to another. One transition of particular importance, because of the large role momentum is likely to play in future world population growth, is a gradual fall in fertility to replacement level.

As discussed in Section 3.3, the population growth associated with gradual declines to stationarity has been analyzed from a number of perspectives. Here we follow Schoen and Jonsson (2003) and focus on how the metastable model can be used to capture that process. The basic approach is to set a value of k such that net maternity moves toward replacement level and, after that level is attained, to hold net maternity at replacement level indefinitely.

Let us assume that the data are in the form of the usual $10 \times 10$ Leslie matrix, which spans the first 50 years of life using ten 5-year age intervals.

The first step is to put that Leslie matrix, $\mathbf{Z_0}$, into the form of matrix $\mathbf{A_t}$ in equation (7.23). Mortality can be removed from the subdiagonal and incorporated in the first row elements. To do so, let $s_j$ be the jth subdiagonal element. Leave the first element of the first row of $\mathbf{Z_0}$ as is, and replace the jth element of the first row, $z_j$, by $s_1\ s_2 \ldots\ s_{j-1}\ z_j$. Then set the subdiagonal elements equal to 1 to establish new matrix $\mathbf{A_0}$. The eigenvectors of the original Leslie matrix change, but the dominant eigenvalue and the future birth trajectory remain the same. The dominant eigenvalue of $\mathbf{Z_0}$, c, can be found using a standard mathematical package (e.g. Maple, Mathematica, or S+). The $f_j$ of the new $\mathbf{A_0}$ matrix can be found from the relationship [adjusted $z_j$] $= f_j\ c^j$, and the procedure verified by confirming that the $f_j$ sum to one.

Assume that the decline from the initial level of the $\mathbf{Z_0}$ matrix to replacement takes N intervals (each of 5 years). At the end of the Nth interval, we want $\mathbf{A_0}$ to be a stationary level matrix identical to matrix $\mathbf{F}$ in equation (7.25). That result can be obtained by setting $k = c^{-1/N}$. Because of the nature of the metastable model, the decline in net maternity over time is exponential, with the largest decreases in the earliest intervals. Although nonlinear, that pattern of decline is quite reasonable.

After the Nth internal, the logic underlying equation (7.31) indicates that the size and age composition of the population are given by

$$\mathbf{x_N} = c^{(N-1)/2}\ \mathbf{S}^N\ \mathbf{U_0}^{-1}\ \mathbf{x_0} \qquad (7.50)$$

as the product of the c and k terms simplify, and $\mathbf{U_N}$ is the identity matrix. From that point on, the behavior of the population is determined by matrix $\mathbf{F}$, and it ultimately becomes stationary. At time point "∞", when $\mathbf{F}$ has become a rank one matrix, that stationary population is given by

$$\mathbf{x_\infty} = \mathbf{c}^{(N-1)/2}\ \mathbf{u_F}\ \mathbf{v_F}'\mathbf{S}^N\ \mathbf{U_0}^{-1}\ \mathbf{x_0} \qquad (7.51)$$

where $\mathbf{u_F}$ is the dominant right eigenvector of $\mathbf{F}$ and $\mathbf{v_F}'$ is its dominant left eigenvector. The first element of $\mathbf{x_\infty}$ gives the size of the ultimate birth cohort, $Q_M$. Momentum, $\Omega_M$, can then be found from equation (3.1) which gives

$$\Omega_M = be(0)Q_M \qquad (7.52)$$

with b the crude birth rate of the initial population and e(0) the ultimate life expectancy.

Equation (7.51) can be simplified considerably by introducing some fairly reasonable assumptions. If the initial population is stable at growth rate c, then $\mathbf{U_0}^{-1}\ \mathbf{x_0} = \mathbf{u_F}$, i.e. a column vector of ones. Even for short declines, $\mathbf{S}^N$ can be approximated by its rank one form, i.e. by $\lambda_s^N \mathbf{u_s v_s}'$ and we have

$$\mathbf{x_\infty} \approx \mathbf{c}^{(N-1)/2}\ \lambda_s^N\ \mathbf{u_F}\ (\mathbf{v_F}'\mathbf{u_s v_s}'\mathbf{u_F}) \qquad (7.53)$$

The product of the 4 vectors in parentheses in equation (7.53) is a scalar. The product of the first two is the change in birth cohort size resulting from the shift from the stable population defined by **S** to the stationary population defined by **F**. The product of the last two vectors yields the long term change in birth cohort size produced by a change from the stationary population defined by **F** to the stable population defined by **S**. Those two adjustments largely offset one another, leaving the product $(\mathbf{v_F'u_s v_S'u_F}) \approx 1$. Under those approximations, the momentum associated with an N interval decline in fertility to replacement, $\Omega^*_M$, becomes

$$\Omega^*_M = b\, e(0)\, c^{(N-1)/2}\, \lambda_s^N \tag{7.54}$$

Equation (7.54) indicates that momentum following a gradual decline to replacement is increased by a factor representing continued growth at the initial stable rate for half the length of the decline, a factor found in previous work. In addition, there is a new factor, $\lambda_s^N$, which is less than 1 when $k < 1$. The $\lambda_s^N$ factor reduces momentum by the negative growth of the **S** matrix over the period of decline. As the age structure of the population moves from its initial stable composition to its ultimate stationary composition, its overall growth is lessened because the falling fertility schedule shifts toward the younger ages, where it characterizes the behavior of relatively smaller cohorts.

Table 7.2 shows momentum values for two hypothetical populations and for fertility declines ranging from 0 to 95 years. The momentum values from equation (7.53) are quite good for delays of 5 years or more, as compared to exact values from a population projection (or equation (7.51)). The values from equation (7.54) are nearly as good, especially for declines of 10 or more years. Both equations (7.53) and (7.54) improve in accuracy as the period of decline lengthens, because their underlying rank 1 assumption becomes closer to reality. It is clear from Table 7.2 that even populations with moderate fertility (an NRR of 1.90) can grow substantially, especially if the decline to replacement level is prolonged. Any such momentum figure must be regarded with some reservations, however. Aside from the inherent difficulties in predicting the future course of fertility, assumptions regarding the age and, especially, the time pattern of decline can exert a substantial influence on the final momentum value.

The approach used to model a transition to stationarity can be applied to modeling the shift from a stable population growing at rate $c_1$ to one growing at rate $c_2$ (Schoen and Jonsson 2003). The transition begins with $\mathbf{A_0}$, a Leslie matrix in the form of matrix $\mathbf{A_t}$ in equation (7.23) that has a dominant eigenvalue of $c_1$. It ends with $\mathbf{B_0}$, a Leslie matrix of the same form with dominant eigenvalue $c_2$. For a decline of N 5-year intervals, the change parameter of the metastable

*Table 7.2. Measures of Population Momentum in Hypothetical Populations Experiencing Exponential Fertility Declines Over Age and Time That Require Y Years to Achieve Replacement Level Fertility*

| | High Fertility Population[a] | | | Moderate Fertility Population[b] | | |
|---|---|---|---|---|---|---|
| Length (Y) | "Actual" (Projection) | Eqn (7.53) | Eqn (7.54) | "Actual" (Projection) | Eqn (7.53) | Eqn (7.54) |
| 0 | 2.04 | 2.10 | 2.00 | 1.59 | 1.61 | 1.58 |
| 5 | 2.22 | 2.22 | 2.18 | 1.68 | 1.68 | 1.67 |
| 10 | 2.45 | 2.44 | 2.42 | 1.77 | 1.77 | 1.77 |
| 15 | 2.71 | 2.71 | 2.69 | 1.87 | 1.87 | 1.87 |
| 20 | 3.01 | 3.01 | 3.00 | 1.98 | 1.98 | 1.98 |
| 25 | 3.36 | 3.36 | 3.35 | 2.10 | 2.10 | 2.10 |
| 30 | 3.75 | 3.75 | 3.74 | 2.23 | 2.23 | 2.23 |
| 35 | 4.19 | 4.18 | 4.18 | 2.36 | 2.36 | 2.36 |
| 40 | 4.67 | 4.67 | 4.67 | 2.51 | 2.51 | 2.51 |
| 45 | 5.22 | 5.22 | 5.22 | 2.66 | 2.66 | 2.66 |
| 50 | 5.83 | 5.83 | 5.83 | 2.82 | 2.82 | 2.82 |
| 55 | 6.52 | 6.52 | 6.52 | 2.99 | 2.99 | 2.99 |
| 65 | 8.14 | 8.14 | 8.14 | 3.36 | 3.36 | 3.36 |
| 75 | 10.17 | 10.17 | 10.17 | 3.78 | 3.78 | 3.78 |
| 85 | 12.71 | 12.71 | 12.71 | 4.25 | 4.25 | 4.25 |
| 95 | 15.88 | 15.88 | 15.88 | 4.78 | 4.78 | 4.78 |

*Notes:* Based on the discrete model with five-year age intervals. Ultimate life expectancy at birth is 75 years. The replacement level age-specific fertility pattern is the "broad peak" pattern of England and Wales Females, 1970 (from Keyfitz and Flieger 1990: 542).

[a] Initial population stable with growth rate $c = 1.250$, $NRR = 3.43$, and birth rate 0.0475.

[b] Initial population stable with growth rate $c = 1.125$, $NRR = 1.90$, and birth rate 0.0290.

*Source:* Adapted from Schoen and Jonsson (2003, Table 1).

transition model is

$$k = (c_2/c_1)^{1/N} \tag{7.55}$$

The jth first row element of $\mathbf{A}_t$, $0 \leq t \leq N$, is $(f_j\ c_1^{\ j}\ k^{jt})$, which is $(f_j\ c_1^{\ j})$ at $t = 0$ and $(f_j\ c_2^{\ j})$ at $t = N$. With $\mathbf{S} = \mathbf{FK}$ and the assumption that $\mathbf{S}^N$ can be written as a rank 1 matrix, the product matrix that projects the initial population from time 0 to time N can be written as

$$\mathbf{M_{0,N}} = c_1^{\ (N-1)/2}\ c_2^{\ (N+1)/2}\ \lambda_s^{\ N}\ \mathbf{U_N}\ \mathbf{u_s v_s{'} U_0}^{-1} \tag{7.56}$$

$\mathbf{U_N}$ is a diagonal matrix containing the relative age composition of $\mathbf{A_N} = \mathbf{B_0}$, whose jth diagonal element is $c_2^{\ -(j-1)}$, and $\mathbf{U_0}$ is the diagonal matrix showing the relative age composition of $\mathbf{A_0}$, with jth diagonal element $c_1^{\ -(j-1)}$.

After N intervals, the population becomes subject to matrix $\mathbf{B_0}$. If we assume that the initial (time 0) population is stable at growth rate $c_1$, and that at time $N+Z$ matrix $\mathbf{B_0}$ can be written in rank 1 form, then the size of the population at time $N+Z$ is

$$\mathbf{x_{N+Z}} = c_1^{\ (N-1)/2}\ c_2^{\ Z+(N+1)/2}\ \lambda_s^{\ N}\ \mathbf{u_B}\ (\mathbf{v_F{'} u_s v_s{'} u_F}) \tag{7.57}$$

as $\mathbf{v_B{'} U_N} = \mathbf{v_F{'}}$ and $\mathbf{U_0}^{-1}\mathbf{u_A} = \mathbf{u_F}$. The subscripts A and B refer to matrices $\mathbf{A_0}$ and $\mathbf{B_0}$ respectively. Equation (7.57) has the same scalar product of four vectors as equation (7.53), and that product can again be approximated as 1. Equation (7.57) indicates that during the transition the population grows at rate $c_1$ for about half the period of decline, and at rate $c_2$ for about half of the period of decline (and for the entire interval after the decline). Once again, there is a factor of $\lambda_s^{\ N}$ that moderates the transitional growth in recognition of the population's changing age structure. The metastable model provides a convenient "bridge" between the two stable regimes, and the resulting projection equations are simplified because the two size adjustment factors yield a product that is close to 1.

The metastable model has the potential to be a valuable analytical tool, as it extends the stable population model that has provided an extremely useful framework for demographic analysis. Metastability is also likely to be helpful in estimating demographic measures in populations with changing rates. The realization of that potential must be left to future work.

## 7.4 INTRINSICALLY DYNAMIC MODELS, AN ALTERNATIVE EXTENSION OF STABILITY

The metastable model provides a complete closed form generalization of the stable model, but is restricted to monotonic increases or decreases in fertility. In this section, we draw on Schoen (2005) and examine a different approach that allows any pattern of fertility change over time, but involves convergent infinite series and a constraining pattern of age-specific fertility change.

An analytically useful dynamic model must provide a solution for the product of a sequence of time-varying Leslie matrices. A general analytical solution is not possible because matrix multiplication does not commute (cf. Gantmacher 1959), but special cases—like metastability—may lead to useful solutions. Another circumstance that simplifies the determination of the birth trajectory is when the population growth rate over a number of intervals (i.e. the dominant eigenvalue of the relevant product matrix) is equal to the product of the growth rates of each interval (i.e. the product of the dominant eigenvalues of the individual interval population projection matrices). Schoen (2005) termed such a model "intrinsically dynamic" because that multiplicative property links the intrinsic growth of each interval with the intrinsic growth across intervals. Tuljapurkar (1990, p83–85) noted that when Leslie matrices shared a common set of reproductive values, overall growth exhibited that multiplicative property. In Leslie matrices without mortality, constant eigenvalues imply constant eigenvectors, so we examine Intrinsically Dynamic Models (IDMs) in the case where only the dominant eigenvalue varies over time.

### 7.4.1 The Two Age Group Intrinsically Dynamic Model

It is easiest to begin with $2 \times 2$ Leslie matrices that ignore mortality. At time t, let the dominant and subordinate roots of IDM matrix $\mathbf{A_t}$ be $\lambda_{1t}$ and (constant) $\lambda_2$. Tuljapurkar (1993 : 265) noted that the elements of a Leslie matrix can be written in terms of its roots, and in the present case that yields

$$\mathbf{A_t} = \begin{bmatrix} \lambda_{1t} + \lambda_2 & -\lambda_{1t}\lambda_2 \\ 1 & 0 \end{bmatrix} \tag{7.58}$$

It is apparent that $\lambda_2 < 0$. Using the relationship following Appendix equation (A.8), it can readily be verified that the (1,1) element of $\mathbf{A_t}$ is $a_t = \lambda_{1t} + \lambda_2$ and the (1,2) element $b_t = -\lambda_{1t}\lambda_2$. It is evident from equation (7.58) that the elements of a $2 \times 2$ Leslie matrix are simple and symmetric functions of both roots.

Consider the product matrix $\mathbf{M_{0,2}} = \mathbf{A_2}\ \mathbf{A_1}$. In terms of its own eigen-structure, equation (2.1) allows that product matrix to be written

$$\mathbf{M_{0,2}} = \mathbf{U_{M2}}\ \mathbf{\Lambda_{M2}}\ \mathbf{V_{M2}} \tag{7.59}$$

Using equations (A.8) through (A.10), we have

$$\mathbf{U_{M2}} = \begin{bmatrix} 1 & 1 \\ \dfrac{\lambda_{11} + \lambda_2}{\lambda_{11}(\lambda_{12} + \lambda_2)} & \dfrac{1}{\lambda_2} \end{bmatrix} \tag{7.60}$$

$$\mathbf{\Lambda}_{\mathbf{M2}} = \begin{bmatrix} \lambda_{12}\lambda_{11} & 0 \\ 0 & \lambda_2{}^2 \end{bmatrix} \tag{7.61}$$

and

$$\mathbf{V}_{\mathbf{M2}} = \begin{bmatrix} 1 & -\lambda_2 \\ \frac{-\lambda_2(\lambda_{11}+\lambda_2)}{\lambda_{11}(\lambda_{12}+\lambda_2)} & \lambda_2 \end{bmatrix} \left\{ \frac{\lambda_{11}(\lambda_{12}+\lambda_2)}{\lambda_{12}\lambda_{11}-\lambda_2^2} \right\} \tag{7.62}$$

That decomposition reveals three important characteristics of IDMs. First, the desired multiplicative property holds, with the $\mathbf{\Lambda}_{\mathbf{M2}}$ matrix indicating that the property applies to both the dominant and subordinate roots. Second, the subordinate right eigenvector $(1/\lambda_2)$ is the same in the product matrix and in both **A** matrices. Third, the relative sizes of the first row elements of the left eigenvector matrices are the same in $\mathbf{M}_{\mathbf{0,2}}$, $\mathbf{A}_{\mathbf{1}}$, and $\mathbf{A}_{\mathbf{2}}$. In $\mathbf{M}_{\mathbf{0,2}}$, however, those elements are multiplied by the time-varying factor in curly brackets to the right of the matrix in equation (7.62). That factor can be interpreted as $1/\mu_{M2}$, the mean age at childbearing under product matrix $\mathbf{M}_{\mathbf{0,2}}$.

The NRR of Leslie matrices with no mortality is the sum of the first row elements. Here, at time t, that is $\lambda_{1t} + \lambda_2 - \lambda_{1t}\lambda_2$. Generally, in terms of elements of **A**, the mean age at childbearing $\mu_A$ is given by

$$\mu_A = a/\lambda_1 + 2b/\lambda_1{}^2 = (\lambda_1 - \lambda_2)/\lambda_1 \tag{7.63}$$

The dominant right eigenvector of the product matrix is not constant over time, allowing the age structure of the population to change. To examine how it changes over time, consider a distant time Z when $\mathbf{M}_{\mathbf{0,Z}}$ is a rank 1 matrix and only the dominant component of the dynamic population is present. The logic of equations (7.59) through (7.62) is that the rank 1 matrix can be written

$$\mathbf{M}_{\mathbf{0,Z}} = \{\prod_{j=1}^{Z} \lambda_{1j}\} \begin{bmatrix} 1/\mu_{MZ} \\ \frac{1}{\lambda_{1Z}(\mu_{M,Z-1})} \end{bmatrix} [1 \quad -\lambda_2] \tag{7.64}$$

where $\mu_{Mt}$ is the mean age at childbearing in product matrix $\mathbf{M}_{\mathbf{0,t}}$ and the equation incorporates the mean ages at childbearing into the column vector so that the row vector is constant. At time Z, the (1,1) element of $\mathbf{M}_{\mathbf{0,Z}}$ is the product of the Z time-varying dominant eigenvalues times $(1/\mu_{MZ})$. The (2,1) element at time Z is the (1,1) element at time Z − 1, that is the product of the first (Z − 1) $\lambda_1$'s times $(1/\mu_{M,Z-1})$. Equation (7.64) thus gives the size and structure of the long term IDM in terms of individual population projection matrix (PPM) eigenvalues and the yet to be determined mean ages $\mu_{Mt}$.

To find the $\mu_{Mt}$, begin by observing that the relative sizes of the elements in the column vector in equation (7.64) and the known constant subordinate eigenvector imply that

$$\mathbf{U}_{Mt} = \begin{bmatrix} 1 & 1 \\ \dfrac{\mu_{Mt}}{\lambda_{1t}(\mu_{M,t-1})} & \dfrac{1}{\lambda_2} \end{bmatrix} \tag{7.65}$$

At any time $\mathbf{V} = \mathbf{U}^{-1}$, so equation (7.65) leads to $\mathbf{V}_{Mt}$ and the determination that its scalar factor is

$$1/\mu_{Mt} = \lambda_{1t}/\{\lambda_{1t} - \lambda_2[\mu_{Mt}/\mu_{M,t-1}]\} \tag{7.66}$$

Equation (7.66) leads to a recursive relationship for $\mu_{Mt}$. Using that relationship at times $t-1$, $t-2$, etc. to eliminate $\mu_M$ values at times before t generates the infinite series solution

$$\frac{1}{\mu_{Mt}} = 1 + \frac{\lambda_2}{\lambda_{1t}} + \frac{\lambda_2{}^2}{\lambda_{1t}\ \lambda_{1,t-1}} + \frac{\lambda_2{}^3}{\lambda_{1t}\ \lambda_{1,t-1}\ \lambda_{1,t-2}} + \frac{\lambda_2{}^4}{\lambda_{1t}\ \lambda_{1,t-1}\ \lambda_{1,t-2}\ \lambda_{1,t-3}} + \cdots \tag{7.67}$$

The series in equation (7.67) must converge because the terms alternate in sign (as $\lambda_2 < 0$) and the absolute value of the ratio $\lambda_2/\lambda_1$ is always less than one because $\lambda_1$ is the dominant root. It can be verified algebraically that the relationships in equations (7.64) and (7.67) satisfy the basic projection equation $\mathbf{x}_t = \mathbf{A}_t\ \mathbf{x}_{t-1}$ when the population vectors are in equation (7.64) form and $\mathbf{A}_t$ is an IDM projection matrix in the form of equation (7.58). In the stable case when the dominant root is constant, the series in equation (7.67) sums to $\lambda_1/(\lambda_1 - \lambda_2)$, the stable solution in equation (7.63).

Equation (7.67) indicates that $\mu_{Mt}$ increases whenever $\lambda_{1t}$ increases. It also describes the process of how, as time passes, the impact of a one time increase in $\lambda_1$ is attenuated by successive movement to smaller and smaller terms in the series. Equation (7.67) is an unavoidable complication because, by adjusting birth cohort size, it is needed to reconcile IDM population growth at $\lambda_{1t}$ with the population's age structure, which bears the imprint of past vital rates.

The two age group IDM solution of equations (7.64) and (7.67) was confirmed by numerical calculations with hypothetical data. With $\lambda_2 = -0.4$, convergence occurred fairly quickly. Six terms of the series in equation (7.67) provided $\mu_{Mt}$ values that were accurate to about 0.002, and nine terms gave values accurate to about 0.0001.

### 7.4.2 The Three Age Group Intrinsically Dynamic Model

In a $3 \times 3$ IDM Leslie matrix with no mortality, the subdiagonal elements are both equal to one and the first row elements are $(\lambda_{1t}+\lambda_2+\lambda_3)$, $-(\lambda_{1t}\lambda_2+\lambda_{1t}\lambda_3+\lambda_2\lambda_3)$, and $(\lambda_{1t}\lambda_2\lambda_3)$. The NRR is again the sum of the first row elements. At time t that yields

$$R_t = 1-(1-\lambda_{1t})(1-\lambda_2)(1-\lambda_3) \qquad (7.68)$$

With three age groups, the time t PPM mean age at childbearing is given by

$$\mu_{At} = (\lambda_{1t}-\lambda_2)(\lambda_{1t}-\lambda_3)/\lambda_{1t}^{\,2} \qquad (7.69)$$

The model specifying approach here parallels that used in the two age group case. The rank 1 product matrix that spans the interval from times 0 to Z can be written

$$\mathbf{M}_{0,Z} = \{\prod_{j=1}^{Z} \lambda_{1j}\} \begin{bmatrix} 1/\mu_{MZ} \\ \dfrac{1}{\lambda_{1Z}(\mu_{M,Z-1})} \\ \dfrac{1}{\lambda_{1Z}\lambda_{1,Z-1}(\mu_{M,Z-2})} \end{bmatrix} [1 \quad -(\lambda_2+\lambda_3) \quad \lambda_2\lambda_3] \qquad (7.70)$$

The third row element of the column vector at time Z relates to the first row element of the corresponding vector at time $Z-2$, i.e. it is $1/(\mu_{M,Z-2})$ divided by the product of the growth rates from $Z-2$ to Z. To find the long term $\mu_{Mt}$, create $\mathbf{U}_{Mt}$ in a manner analogous to equation (7.65), invert it to find the factor multiplying $\mathbf{V}_{Mt}$, and use that factor recursively to obtain an equation similar to equation (7.67). The result is

$$\frac{1}{\mu_{Mt}} = 1+\frac{\lambda_2+\lambda_3}{\lambda_{1t}}+\frac{(\lambda_2^{\,3}-\lambda_3^{\,3})/(\lambda_2-\lambda_3)}{\lambda_{1t}\,\lambda_{1,t-1}}+\frac{(\lambda_2^{\,4}-\lambda_3^{\,4})/(\lambda_2-\lambda_3)}{\lambda_{1t}\,\lambda_{1,t-1}\,\lambda_{1,t-2}}$$
$$+\frac{(\lambda_2^{\,5}-\lambda_3^{\,5})/(\lambda_2-\lambda_3)}{\lambda_{1t}\,\lambda_{1,t-1}\,\lambda_{1,t-2}\,\lambda_{1,t-3}}+\cdots \qquad (7.71)$$

Equations (7.70) and (7.71) provide the solution for the 3 age group IDM. The infinite series in equation (7.71) converges because it is an alternating series and $\lambda_{1t} \geq -(\lambda_2+\lambda_3)$ since $(\lambda_{1t}+\lambda_2+\lambda_3) = a_t \geq 0$. Equations (7.70) and (7.71) satisfy the basic projection relationship algebraically, regardless of whether the subordinate roots are real and unequal, real and equal, or complex conjugates. Calculations using hypothetical data have verified those relationships numerically. Convergence of the series for $\mu_{Mt}$ is a bit slower than in the 2 age group case, with accuracy to about 0.01 to 0.02 after 7 terms and about 0.001 after 10 terms.

Table 7.3 considers four different fertility patterns and how age-specific fertility varies with the value of the dominant root. The two subordinate roots set the age pattern, and the four patterns were chosen to span the range of commonly encountered three age group discrete fertility schedules. Pattern 1 emphasizes early fertility, Pattern 2 mid-reproductive age fertility, and Patterns 3 and 4 late fertility. Patterns 1 and 4 have complex conjugate subordinate roots, Pattern 2 has two real and unequal subordinate roots, and Pattern 3 has a double subordinate root.

In every pattern, at every age, fertility varies linearly with $\lambda_1$. When $\lambda_1$ increases, age group 0–14 always has the largest increase, age group 15–29 the second largest, and age group 30–44 the smallest. That pattern is consistent with the inverse relationship between $\lambda_1$ and the mean age at childbearing. Those fertility changes depart from proportionality, but even if not typical they are plausible given the variability in human fertility. Nonetheless, Pattern 1 has atypically high relative fertility at ages 0–14 when $\lambda_1$ is 1.5 (and Lotka's $r = 0.027$), and Patterns 3 and 4 have zero fertility in that age group when $\lambda_1$ is 0.8 (and Lotka's $r = -0.015$). Since the size of the first age group moves in lockstep with $\lambda_1$, values of $\lambda_1$ less that 0.8 would produce unacceptable (negative) values.

*Table 7.3. Age-Specific Fertility Patterns and their Variation With the Dominant Root in Three Age Group Intrinsically Dynamic Models*

| | Pattern 1 | | | Pattern 2 | | | Pattern 3 | | | Pattern 4 | | |
|---|---|---|---|---|---|---|---|---|---|---|---|---|
| Root ($\lambda_1$) | a | b | c | a | b | c | a | b | c | a | b | c |
| 0.8 | .2 | .38 | .08 | .1 | .46 | .08 | 0 | .48 | .128 | 0 | .44 | .16 |
| 0.9 | .3 | .44 | .09 | .2 | .53 | .09 | .1 | .56 | .144 | .1 | .52 | .18 |
| 1.0 | .4 | .5 | .1 | .3 | .6 | .1 | .2 | .64 | .16 | .2 | .6 | .2 |
| 1.1 | .5 | .56 | .11 | .4 | .67 | .11 | .3 | .72 | .176 | .3 | .68 | .22 |
| 1.2 | .6 | .62 | .12 | .5 | .74 | .12 | .4 | .8 | .192 | .4 | .76 | .24 |
| 1.3 | .7 | .68 | .13 | .6 | .81 | .13 | .5 | .88 | .208 | .5 | .84 | .26 |
| 1.4 | .8 | .74 | .14 | .7 | .88 | .14 | .6 | .96 | .224 | .6 | .92 | .28 |
| 1.5 | .9 | .8 | .15 | .8 | .95 | .15 | .7 | 1.04 | .24 | .7 | 1.0 | .3 |
| Subordinate Roots | $-0.3 \pm .1i$ | | | $-0.2, -0.5$ | | | $-0.4$ (double root) | | | $-0.4 \pm .2i$ | | |

*Note:* Column headings a, b, and c refer, respectively, to the first, second, and third elements of the first row of the Leslie matrix.
*Source:* Schoen (2005).

The constant subordinate eigenstructure of the IDM model was adopted for its mathematical, not its demographic properties. By restricting changes in the age pattern of fertility, it allows an analytical determination of the birth trajectory generated by any sequence of NRRs, and provides a consistent set of underlying age-specific fertility values. Although Table 7.3 indicates that care must be taken to avoid negative values, the IDM model performs reasonably well with 15-year age groups. It relaxes the constant rate stable restriction and allows much greater flexibility in the nature of fertility change than does the metastable model.

### 7.4.3 The n-Age Group Intrinsically Dynamic Model

The equations for the case of n age groups can be derived by extending the previous approach. The Tuljapurkar (1993: 265) relationship for the first row elements of each Leslie matrix is essentially that the first term of the first row is the sum of the roots, the second term is minus the sum of the roots taken two at a time, the third term is the sum of the roots taken three at a time, and so on. Thus the jth first row element at time t, $a_{jt}$, is $(-1)^{j+1}$ times the sum of all possible combinations of the roots taken j at a time. All of the first row elements of $\mathbf{A}_t$ are linear functions of $\lambda_{1t}$.

The NRR in the n-age group, no mortality case can be expressed algebraically as

$$R_t = 1 - \prod_{j=1}^{n} (1 - \lambda_{jt}) \tag{7.72}$$

The mean age at childbearing implied by PPM $\mathbf{A}_t$ is given by

$$\mu_{At} = a_{1t}/\lambda_{1t} + 2a_{2t}/\lambda_{1t}^{2} + 3a_{3t}/\lambda_{1t}^{3} + \ldots + na_{nt}/\lambda_{1t}^{n} \tag{7.73}$$

In terms of the roots, that relationship can be written

$$\mu_{At} = \{\lambda_{1t}^{1-n}\} \prod_{j=2}^{n} (\lambda_{1t} - \lambda_{j}) \tag{7.74}$$

The product matrix in the n age group case is an extension of equation (7.70), its form in the three age group case. A recursive equation for $\mu_{Mt}$ can again be derived, though it becomes increasingly complicated as the number of age groups increases.

There is a serious problem with the IDM approach when $n > 3$, not because of complexity but because of the pattern of change in the age-specific rates. Consider the conventional $10 \times 10$ Leslie model with 5-year age groups.

Since human populations have essentially zero fertility under age 10, the (1,1) element of every $10 \times 10$ Leslie matrix must be zero. However, the (1,1) element is the sum of all of the roots. If that sum is zero at one time point, any change in the dominant root would give it a demographically unacceptable value. The problem is not so stark when n is 4 or 5, but the pattern of change can be unrealistic.

The $3 \times 3$ model emerges as the most useful IDM in the birth-death (multi-age) case. Patterns of fertility are usually acceptable, and 15-year age intervals coincide quite well with the beginning of childbearing, the mean age at childbearing, and the end of childbearing. When 5-year age data are available, the observed $10 \times 10$ Leslie matrix can readily be condensed to $3 \times 3$ matrix. A procedure for doing so is set forth in Keyfitz (1968, 37–40).

### 7.4.4 IDM Birth Trajectories

The trajectory of IDM births over time is, apart from adjustments reflecting changes in the product matrix mean age at childbearing, determined by the product of the dominant roots of the individual PPMs. When the $\lambda_{1t}$ follow a known functional form, it is often possible to find their product analytically. If $\lambda_{1t} = \exp[ht]$ for some known constant h, the usual summation formula from 1 to t can be applied to yield

$$\lambda_{Mt} = \exp[h\ t\ (t+1)/2] \tag{7.75}$$

an exponentiated quadratic trajectory. If $\lambda_{1t} = \exp[ht^2]$, then the standard summation relationship gives

$$\lambda_{Mt} = \exp[ht\ (t+1)(2t+1)/6] \tag{7.76}$$

and if $\lambda_{1t} = \exp[ht^3]$, we have

$$\lambda_{Mt} = \exp[ht^2(t+1)^2/4] \tag{7.77}$$

The case where $\lambda_{1t}$ varies cyclically is of particular interest, and Schoen (2005) found that if $\lambda_{1t} = \exp[h\ \sin(\omega t)]$, then

$$\lambda_{Mt} = \exp[(h/2)\{\sin(\omega) - \sin(\omega[t+1]) + [\cot(\omega/2)][\sin(\omega)\cos(\omega) - \cos(\omega[t+1])]\}] \tag{7.78}$$

Equation (7.78) resembles equation (7.11), the Constant Generation Length relationship for the birth trajectory resulting from a cyclical NRR sequence. As an exponentiated polynomial change in fertility levels leads to population growth according to an exponentiated polynomial one degree higher, sinusoidal change in fertility appears to lead to population cycles that follow a trajectory determined by a combination of sine and cosine functions.

### 7.4.5 Observed Populations as IDM Populations

Any given observed population can be viewed as an IDM population, although that IDM representation is not unique. Schoen (2005) derived a representation for the case of a time 0 reproductive population with three age groups, assuming that the population was stable with dominant root $\lambda_A$ and subordinate roots $\lambda_2$ and $\lambda_3$ before time $(-1)$. With $\mathbf{x_0}$, the given observed time 0 population, scaled so that there is one person in the first age group, $x_{20}$ persons in the second age group, and $x_{30}$ persons in the third age group, that IDM representation can be written

$$\mathbf{x_0} = \begin{bmatrix} 1 \\ [\lambda_A + \mu_A(\lambda_{M0} - \lambda_A)]^{-1} \\ [\lambda_A{}^2 + \lambda_A\ \mu_A(\lambda_{M0} - \lambda_A)]^{-1} \end{bmatrix} \tag{7.79}$$

where $\lambda_A = x_{20}/x_{30}$, $\mu_A = (\lambda_A - \lambda_2)(\lambda_A - \lambda_3)\lambda_A{}^{-2}$, and $\lambda_{M0} = [1 + x_{20}\ \lambda_A \times (\mu_A - 1)]/(x_{20}\ \mu_A)$. The age composition of the population is stable at $\lambda_A$ when $\lambda_A = \lambda_{M0}$, i.e. when $x_{20}/x_{30} = 1/x_{20}$.

### 7.4.6 Transitions and Momentum in IDM Populations

Transitions from one set of fixed rates to another can readily be modeled using IDMs. Assume that a stable population with dominant root $\lambda_A$ shifts to a new vital rate regime with dominant root $\lambda_B$, the subordinate roots remaining constant. We consider a three age group population with the number of persons in the first age group scaled to 1 at $t = 0$. With the shift in rates occurring immediately after time 0, equations (7.70) and (7.71) describe the population trajectory. Figure 7.3 shows the birth trajectory (i.e. the number of persons age 0–14) when $\lambda_A = 1.5$ and $\lambda_B = 1$ under Patterns 2 and 4 of Table 7.3. After fertility falls to replacement, the number of births drops sharply, partially recovers, and then continues to oscillate with fluctuations of diminishing amplitude around the ultimate stationary level. Figure 7.4 shows that the product matrix mean age at childbearing, $\mu_{Mt}$, follows a rather similar path as it moves from $\mu_A$ to $\mu_B$.

The population momentum associated with such a change in vital rates can be found by applying equation (3.1). In an IDM, equation (7.70) indicates that the size of the ultimate stationary birth cohort relative to the initial birth cohort, Q, is simply

$$Q = \mu_A/\mu_B \tag{7.80}$$

*Figure 7.3. The IDM Birth Trajectory Following a Transition, at Time 0, From a Stable ($\lambda_A = 1.5$) to a Stationary ($\lambda_B = 1$) Regime, Under Fertility Patterns 2 and 4*

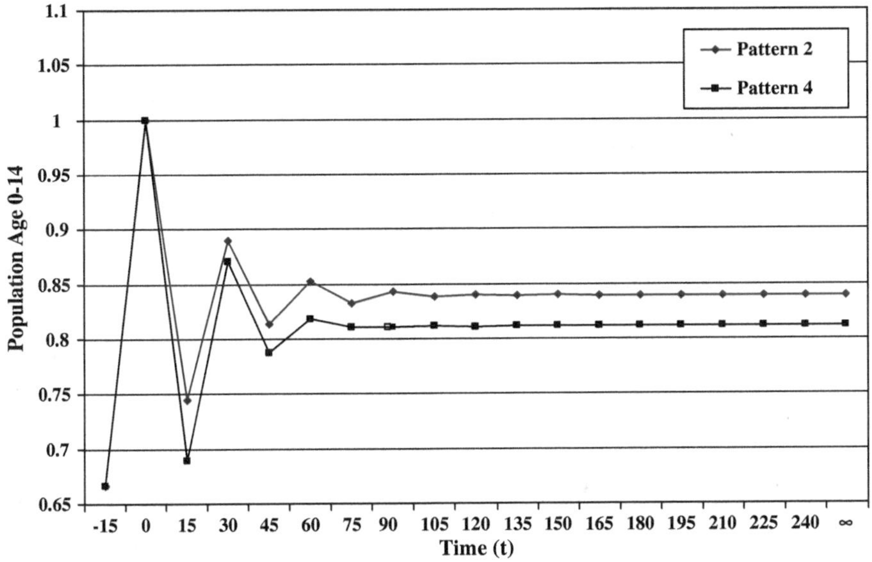


*Source:* Schoen (2005).

Under Pattern 2, with $\lambda_A = 1.5$ and $\lambda_B = 1$, that ratio is $22\frac{2}{3}/27$ or 0.8395. Under Pattern 4, the ratio is $24^1/_3/30$ or 0.8111. In both cases, the fall in fertility leads to an increase in the mean age at childbearing.

In general, birth cohort size adjustment Q can be applied to the change in relative cohort size associated with any change in vital rates, after removing the effects of changes in overall population growth. From equation (7.70), the ratio from time $\alpha$ to time $\beta$ is

$$Q = \mu_{M\alpha}/\mu_{M\beta} \tag{7.81}$$

The IDM approach can thus examine population growth associated with any given path to stationarity, expanding the metastable analysis in subsection 7.3.3. Let the initial population mean age at childbearing be $\mu_{M0}$, which can always be found via equation (7.79). With ultimate stationary mean age at childbearing $\mu_B$ at time $\beta$, and $\lambda_{1t}$, $0 < t < \beta$, the PPM growth rate at time t, the relative size of the ultimate birth cohort, $Q^*$, is

$$Q^* = \{\prod \lambda_{1j}\}\ \mu_{M0}/\mu_B \tag{7.82}$$

*Figure 7.4. The IDM Cumulative Mean Age of Childbearing ($\mu_{Mt}$) Following a Transition, at Time 0, From a Stable ($\lambda_A = 1.5$) to a Stationary ($\lambda_B = 1$) Regime, Under Fertility Patterns 2 and 4*

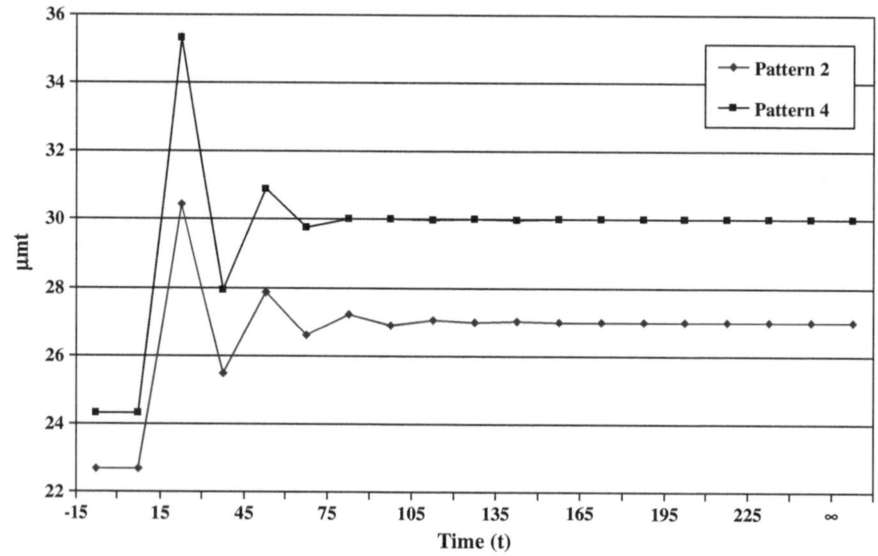


*Source:* Schoen (2005).

where the product is taken from j = 1 to j = B. The associated momentum follows immediately from equation (3.1). The relationship is exact in IDMs, i.e. when the multiplicative property links the λs of the PPMs and the product matrices. In hypothetical calculations, Schoen (2005) found that equation (7.82) gave results that were quite close to projections assuming proportional declines in age-specific fertility rather than IDM pattern fertility declines.

## 7.5 FINDING RATES CONSISTENT WITH A KNOWN BIRTH TRAJECTORY

Instead of seeking the birth trajectory implied by a given sequence of vital rates, the problem can be approached from the other direction, i.e. one can seek a sequence of Leslie matrices that yields a given birth trajectory. There is no unique solution, as any birth trajectory can be produced by many different PPM sequences, each following from different assumptions.

### 7.5.1 Hyperstable Models

Following Schoen and Kim (1994a), consider population renewal equation (7.2) under the assumption that persons at each age x have a fixed proportion of all births at every time, i.e. that

$$f(x) = [g(t-x)/g(t)]\ \phi(x,t) \tag{7.83}$$

where f(x) is the time invariant contribution from age x, with $\int f(x)dx = 1$. That assumption usually yields demographically plausible values (Schoen and Kim 1994a), and leads to the array of net maternity rates

$$\phi(x,t) = f(x)\ [g(t)/g(t-x)] \tag{7.84}$$

or, in terms of known net maternity pattern $\phi(x,0)$

$$\phi(x,t) = \phi(x,0)\ [g(t)g(-x)]/g(t-x) \tag{7.85}$$

That "fixed f" or "hyperstable" assumption immediately links any birth trajectory g(t) with a consistent set of net maternity rates. To be specific, let $g(t) = \exp[ht^2/2]$. Then from equation (7.85) we can write

$$\phi(x,t) = \phi(x,0)\ \exp[hxt] \tag{7.86}$$

the same relationship as in metastable equation (7.34). From equation (7.2), the renewal or characteristic equation of the model is then

$$1 = \int \exp[hx^2/2]\ \phi(x,0)\ dx \tag{7.87}$$

Metastable models are closely related to exponentially quadratic hyperstable models, but the two are not identical. Exercise 12 shows that with the identical birth trajectory, the two models have different sets of vital rates.

If the birth trajectory is the cubic relationship $g(t) = \exp[ht^3]$, net maternity is given by

$$\phi(x,t) = \phi(x,0)\ \exp[3hxt(t-x)] \tag{7.88}$$

and the hyperstable renewal equation is

$$1 = \int \exp[-hx^3]\ \phi(x,0)\ dx \tag{7.89}$$

In the general polynomial form

$$g(t) = \exp[\Sigma\ h_j\ t^j] \tag{7.90}$$

where the sum runs from j = 1 to j = n, net maternity is

$$\phi(x, t) = \phi(x, 0)\ \exp[\Sigma\ h_j\{t^j + (-x)^j - (t - x)^j\}] \qquad (7.91)$$

and the hyperstable characteristic equation is

$$1 = \int \exp[\Sigma h_j(-x)^j]\ \phi(x, 0)\ dx \qquad (7.92)$$

Sinusoidal and other birth functions can easily be incorporated as well (e.g. see Exercise 9). The stable population is the special case where $g(t) = \exp[rt]$. Once established, of course, the birth trajectory/vital rate relationship can be pursued in either direction.

It is useful to write the hyperstable model in discrete form. At time t, let $n \times n$ hyperstable PPM $\mathbf{A_t}$ have the form

$$\mathbf{A_t} = \begin{bmatrix} f_1 g_t/g_{t-1} & f_2 g_t/g_{t-2} & f_3 g_t/g_{t-3} & \cdots & f_n g_t/g_{t-n} \\ 1 & 0 & 0 & \cdots & 0 \\ 0 & 1 & 0 & \cdots & 0 \\ 0 & 0 & \cdots & 1 & 0 \end{bmatrix} \qquad (7.93)$$

where $f_j$ represents the fixed proportion of births arising from persons in age group j and $g_t$ represents the number of births (persons in the first age group) at time t. As equation (7.93) ignores mortality, PPM $\mathbf{A_t}$ projects the birth sequence. Now let $\mathbf{G_t}$ be an $n \times n$ diagonal matrix whose jth diagonal element is $g_{t-j+1}$. PPM $\mathbf{A_t}$ can then be written

$$\mathbf{A_t} = \mathbf{G_t}\ \mathbf{F}\ \mathbf{G_{t-1}}^{-1} \qquad (7.94)$$

where $\mathbf{F}$ is the matrix defined in equation (7.25). Equation (7.94) provides the hyperstable PPM at any time in terms of the known birth trajectory and the fixed proportional distribution of births. It follows that the product of the $\mathbf{A_t}$ from time 1 to time t is

$$\mathbf{M_{0,t}} = \mathbf{G_t}\ \mathbf{F^t}\ \mathbf{G_0}^{-1} \qquad (7.95)$$

The form of equation (7.95) assures convergence to hyperstability. The contribution of any initial population to the size of the ultimate hyperstable population can be expressed as a constant factor in terms of the dominant left eigenvector of $\mathbf{F}$. Kim (1987) described population dynamics in terms of "forward" and "backward" growth rates that generally differed from each other. The hyperstable population is a special case where both the forward and backward growth rates are equal to the growth rate of births.

### 7.5.2 Models With Proportional Rates

Vital rates consistent with a given birth trajectory can be found by assuming that a known age-specific schedule of net maternity rates varies proportionally at all ages. That assumption assures a reasonable age pattern of rates.

Kim and Schoen (1996a) analyzed models with proportional rates and sinusoidal birth trajectories. Drawing on that work, let $\phi(x)$ be the base schedule of net maternity, scaled so that $\int \phi(x)\ dx = 1$. Then assume that

$$\phi(x, t) = R(t)\ \phi(x) \tag{7.96}$$

or that age-specific net maternity varies directly with the period NRR. Equation (7.2) yields the relationship

$$R(t) = g(t)/[\int g(t-x)\ \phi(x)\ dx] \tag{7.97}$$

Assume that the birth function is given by

$$g(t) = 1 + b\ \sin(\omega t) \tag{7.98}$$

To find a closed form expression for R(t), Kim and Schoen (1996a) assumed that age-specific net maternity was normally distributed with mean $\mu$ and standard deviation $\sigma$. Equations (7.97) and (7.98) then yield

$$R(t) = [1 + b\ \sin(\omega t)]/[1 + \alpha(\omega, \sigma)\ b\ \sin(\omega\{t-\mu\})] \tag{7.99}$$

where $\alpha(\omega, \sigma) = \exp[-\omega^2\sigma^2/2]$. Equation (7.99) shows that the NRR cycles with the same cycle length as g(t), even though it is not a simple sine function.

## 7.6 SUMMARY

Birth-death models with changing vital rates provide a platform for analyzing the dynamics of observed populations and exploring their underlying regularities. This chapter emphasizes relationships between changing rates and the birth trajectories associated with them. Assuming a constant generation length in models that do not explicitly recognize age reveals several overall patterns. If fertility changes according to an exponentiated nth degree polynomial, the associated birth trajectory is an exponentiated polynomial of degree $n+1$. If fertility varies sinusoidally, births vary as a (not necessarily simple) function of sines and cosines with the same cycle length. Model dynamics are largely determined by the ratio of cycle length to generation length. Population age structure, and the intergenerational transfer environment encountered by different cohorts, can vary substantially by position in the cycle.

In dynamic models incorporating age, a key problem is to analytically determine the product matrix in terms of the individual Leslie matrices. Two distinct solutions to that problem are examined. The metastable model provides a full, closed form generalization of the stable model, is based on a realistic pattern of age-specific fertility change, and facilitates analyses of fertility transitions. However, change in the metastable model is limited to monotonic increases or decreases in fertility. Intrinsically Dynamic Models (IDMs) also generalize the stable model, and allow any pattern of change in fertility over time. To do so, the IDM solution requires convergent infinite series, and the age-specific fertility pattern implied by its assumption of constant subordinate roots can be unrealistic for age intervals of less than 15 years. Both approaches contribute to the analysis of population momentum under gradual declines to replacement level reproduction.

Many approaches can yield sequences of Leslie matrices that are consistent with a given birth trajectory, as there is no unique solution. Hyperstable models assume a fixed proportional distribution of births and generate demographically reasonable Leslie matrices for a broad range of birth trajectories. They are readily expressed in discrete form, and the hyperstable model with an exponentiated quadratic birth trajectory is closely related to the metastable model. An alternative approach assumes that fertility changes proportionally at all ages. In the sinusoidal case, the proportional fertility assumption exemplifies the connection between sinusoidal change in births and levels of fertility that are trigonometric functions with the same cycle length.

## 7.7 EXERCISES

[See Appendix B for selected answers]

1. Equations (7.10) and (7.11) can be seen as the solution of a pair of differential equations (Schoen and Kim 1997: 287). If $x = \ln g(t)$ and $y = \ln R(t)$, show that with phase difference $\theta \neq 0$

$$y' = \omega[y \tan(\omega A/2) - 2x \tan(\omega A/2) + b]$$

$$x' = (\omega/2)[y \tan(\omega A/2)\{1 + \cot^2(\omega A/2)\} - 2x \tan(\omega A/2) + b]$$

and that when $\theta = 0$ (e.g. when $T = 2A$) that

$$y' = \pm b\omega[1 - (2x/b)^2]$$

$$x' = \pm b\omega[1 - (y/b)^2]/2$$

where the $\pm$ indicates that x and y always move in the same direction. Verify that the above equations provide a formal parallel between cyclical reproduction models and predator-prey models.

2. For synthesis matrix **S** in equation (7.28), with dominant root $\lambda_s$, verify that
   a. dominant right eigenvector $\mathbf{u}_s$ has first element 1 and jth element $(\lambda_s)^{-(j-1)}\ k^{(j-1)(j-2)/2}$
   b. the ith element of dominant left eigenvector $\mathbf{v}_s$ is given by

$$[\sum_{j=i}^{n} \{\prod_{h=i}^{j-1} k^{(h-1)}\}\ f_j\ k^{(j-1)}\ (\lambda_s)^{(i-j-1)}]/\mu_s$$

   where mean age at childbearing $\mu_s = \Sigma_j\ j\ f_j\ k^{(j-1)/2}\ (\lambda_s)^{-j}$

3. In matrix **S** of equation (7.28), $\lambda_s$ and k are clearly related.
   a. In a $3 \times 3$ **S** matrix, if $f_1 = k^{1/2} - k^2$, $f_2 = k^2$, and $f_3 = 1 - k^{1/2}$, show that $\lambda_s = k^{1/2}$.
   b. In a $10 \times 10$ **S** matrix, evaluate the approximation $\lambda_s \approx k^{2.5}$.

4. Consider a metastable model with $\lambda_s = k^{2.5}$, $k = e^b$ $(b < 0)$, and birth trajectory

$$g(t) = g(0)\ e^{at}\ e^{2.5bt}\ e^{bt(t+1)/2}$$

   Given the Gaussian (normal curve) density distribution $Z(t) = (1/[\sigma\{2\pi\}^{1/2}])\ \exp[-t^2/(2\sigma^2)]$, show that g(t) follows a Gaussian distribution with

$$g(0) = \{-b/(2\pi)\}^{1/2},\ a = -3b,\ \text{and}\ \sigma^2 = -1/b$$

5. Consider the metastable/associated stable population birth crossover point $(B_4)$, i.e. the age at which the metastable and associated stable populations have the same proportion of their births. Using the notation of equation (7.36), show that at every time point

$$B_4 = 2s/h$$

6. Show that in a metastable population where p(x,t) is the probability of surviving to age x at time t, c(x,t), the proportion age x at time t, can be written

$$c(x, t) = b_M(t) \exp[-hxt - x(r+s) + hx^2/2]\ p(x, t)$$

   where $b_M(t)$ is the metastable birth rate at time t.

7. Consider an $n \times n$ Leslie matrix with ones on the subdiagonal and jth first row element $f_j = \lambda^j$ times binomial coefficient $(n-1, j-1)/2^{n-1}$. [Binomial coefficient $(n, j) = n!/\{j!(n-j)!\}$] Show that the NRR is then given by

$$R(t) = \lambda(1+\lambda)^{n-1}/2^{n-1}$$

8. Consider the IDM populations in Figure 7.3. For Patterns 2 and 4
   a. find the initial and final number of persons under age 30.
   b. how do the numbers in part a relate to the claim that the number of persons under age 30 is essentially constant over the transition to stationarity?
9. Show that under hyperstability, if $g(t) = 1 + b\sin(2\pi t/T)$, the net maternity function is given by

$$\phi(x, t) = \phi(x, 0)[1 + b\sin(2\pi t/T)][1 - b\sin(2\pi x/T)]/ \\ [1 + b\sin(2\pi\{t-x\}/T)]$$

and the hyperstable characteristic equation by

$$1 = \int[1 - b\ \sin(2\pi x/T)]\ \phi(x, 0)\ dx$$

10. Let PPM $\mathbf{A}_t = \mathbf{K}^t\ \mathbf{G}_t\ \mathbf{F}\ \mathbf{G}_{t-1}{}^{-1}\ \mathbf{K}^{-t}$ for diagonal matrices $\mathbf{K}$ and $\mathbf{G}$, and matrix $\mathbf{F}$ of equation (7.25). Show that the product matrix for times 1 through Z is given by

$$\mathbf{M}_{0,Z} = \mathbf{K}^Z\ \mathbf{G}_Z\ (\mathbf{F}\ \mathbf{K}^{-1})^Z\ \mathbf{G}_0{}^{-1}$$

11. Consider the birth trajectory $g(t) = \exp[ht^\gamma]$. If $h > 0$ and $0 < \gamma < 1$, show that
    a. $g(t) > 1$ and increasing over time
    b. $R(t) > 1$ and decreasing to 1 over time
12. Consider metastable and hyperstable models where $g(t) = c^t\ k^{t(t+1)/2}\ (\lambda_s)^t$. Show that
    a. the jth first row element of the hyperstable $\mathbf{A}_t$ is $f_j\ c^j\ k^{jt-j(j-1)/2}\ (\lambda_s)^j$
    b. the jth first row element of the metastable $\mathbf{A}_t$ is $f_j\ c^j\ k^{jt}$
    c. that the metastable and hyperstable first row values must be different
13. Consider a $3 \times 3$ Leslie matrix with ones as subdiagonal elements and first row elements a, b, and c. Show
    a. the characteristic (eigenvalue defining) equation of that matrix is

$$\lambda^3 - a\lambda^2 - b\lambda - c = 0$$

b. the roots of that equation are

$$\lambda_1 = [2a + Z + (12b + 4a^2)/Z]/6$$

$$\lambda_2 = [4a - Z - (12b + 4a^2)/Z]/12 + (3^{1/2}i/12)[Z - (12b + 4a^2)/Z]$$

$$\lambda_3 = [4a - Z - (12b + 4a^2)/Z]/12 - (3^{1/2}i/12)[Z - (12b + 4a^2)/Z]$$

where $i$ is the square root of minus one and

$$Z = \{36ab + 108c + 8a^3 + 12[12a^3c + 81c^2 + 54abc - 3a^2b^2 - 12b^3]^{1/2}\}^{1/3}$$

14. Consider hyperstable equation (7.94). Show that in the long term, the time t PPM is unaffected if matrix **F** changes over time, as long as it remains row stochastic.
15. In hyperstable equation (7.94), let the jth diagonal element of matrix **G** be the current number of persons in the jth age group, with the number of persons in the first age group always equal to 1. Show that specifies a hyperstable dynamic mortality model.

# CHAPTER 8

# DYNAMIC MULTISTATE MODELS

## 8.1 INTRODUCTION

In multistate models, changing rates of birth, death, and interstate transfer determine the evolving size and state composition of the population. Since state and age are conceptually distinct, multistate models differ from the multi-age models of the previous chapter, with the differences introducing both complexity and analytical flexibility. Transfers to a variety of states may or may not occur, and because movement is not directly linked to time, discrete models with rates as well as survivorship probabilities are possible.

Multistate life tables have become standard and widely used tools in demography (cf. section 1.6). Although contributing to a number of valuable analyses (e.g. Rogers and Willekens, 1986), the multistate stable population has received less attention (see section 1.7). In part that is because of the strong assumption of fixed rates and the weak connection between observed and multistate stable population compositions. Dynamic multistate models have the potential to overcome that difficulty, but research is still at a relatively early stage. In this chapter we discuss a number of approaches that have been advanced, their underlying assumptions, and the relationships they imply.

We begin with a cohort analysis where time and age are interchangeable, and consider circumstances where it can be assumed that all rates change proportionally over time. Next we examine how the metastable and IDM solutions of Chapter 7 can be applied to find multistate trajectories from specified changes in growth and interstate transfer rates. Multistate hyperstable models, including their specification in terms of rates and their relationship to birth-death models, are then explored. Finally we consider two approaches to estimating multistate rates from known population distributions, noting their conceptual differences and numerical similarities.

## 8.2 COHORT MULTISTATE MODELS WITH PROPORTIONAL TRANSFER RATES

In the basic life table, it is simple to aggregate survivorship over age intervals. The number of survivors to any age is readily found using the discrete version of equation (1.5), where the integral of the forces of mortality is replaced by the sum of the age-specific death rates. In a multistate life table, however, movements between states introduce major complications. As equation (1.63) indicates, multistate survivorship across age intervals requires matrix multiplication, not simply summation. One can always carry out the matrix multiplications involved and obtain numerical solutions, but there is no generally applicable analytical procedure for finding the product of a sequence of matrices. Although multistate models are frequently used in demographic analyses, there are no standard techniques for analyzing such questions as how decreases in fertility levels affect a cohort's parity distribution, how changes in marriage and divorce rates impact a cohort's marital status composition, or how changes in movements between health states influence cohort longevity.

Schoen and Canudas-Romo (2006) advanced a Proportional Transfer Rate (PTR) approach for aggregating effects over age intervals, assuming that all age-state-specific rates of transfer are proportional over age. They applied the model to the analysis of cohort parity progression, and Schoen and Canudas-Romo (2005b) used the same approach to examine a three living state model of first marriage, divorce, and remarriage. This section draws largely on that work.

### 8.2.1 Relationships in Proportional Transfer Rate (PTR) Models

The PTR approach begins with base rate matrix $\mu$. Consider the general case with k living states, and let $\mu$ be a $k \times k$ matrix whose row i and column j element ($i \neq j$) is $\mu_{ji}$, the base rate of movement from state j to state i. The jth diagonal element of $\mu$ is $-\Sigma\mu_{ji}$, where the sum over i includes all states except j. For example, when $k = 3$, the base rate matrix can be written

$$\mu = \begin{bmatrix} -\mu_{12} - \mu_{13} - \mu_{1\delta} & \mu_{21} & \mu_{31} \\ \mu_{12} & -\mu_{21} - \mu_{23} - \mu_{2\delta} & \mu_{32} \\ \mu_{13} & \mu_{23} & -\mu_{31} - \mu_{32} - \mu_{3\delta} \end{bmatrix} \tag{8.1}$$

with $\delta$ representing an absorbing state (usually death). Every age-specific transfer rate matrix is a scalar multiple of base matrix $\mu$, with z(x,n) the factor that sets the level of the transfer rates for the (x,n) age interval. The transfer rate matrix for the (x,n) age interval, $\mathbf{M(x, n)}$, is then

$$\mathbf{M(x, n)} = z(x, n)\boldsymbol{\mu} \tag{8.2}$$

As a result, every $\mathbf{M}$ matrix has the same ratio between every pair of transfer rates, e.g.

$$M_{21}(x, n)/M_{12}(x, n) = M_{21}(x+n, n)/M_{12}(x+n, n) = \mu_{21}/\mu_{12} \tag{8.3}$$

Assuming constant rates within age intervals, we can use the standard exponential solution to transform a transfer rate matrix into a survivorship (or cohort projection) matrix. Paralleling constant forces equation (1.63)

$$\mathbf{A}(\mathbf{x}, \mathbf{n}) = \exp[\mathrm{n}\ \mathbf{M}(\mathbf{x}, \mathbf{n})] = \exp[\mathrm{n}\ \mathrm{z}(\mathrm{x}, \mathrm{n})\boldsymbol{\mu}] \tag{8.4}$$

where **A(x, n)** is the matrix of transition probabilities from exact age x to exact age x + n. The (i,j)th element of **A(x, n)** gives the probability that a person in state j at age x will be in state i at age x + n.

Under the PTR assumption, the rate matrices commute and can readily be aggregated over age. To survive the cohort from age x to age x + w, we form product matrix **P(x, w)**, which can be written

$$\begin{aligned}\mathbf{P}(\mathbf{x}, \mathbf{w}) &= \mathbf{A}(\mathbf{x}+\mathbf{w}-\mathbf{n}, \mathbf{n})\mathbf{A}(\mathbf{x}+\mathbf{w}-2\mathbf{n}, \mathbf{n})\ldots\mathbf{A}(\mathbf{x}, \mathbf{n}) \\ &= \exp[\mathrm{n}\ \Sigma\ \mathrm{z}(\mathrm{s}, \mathrm{n})\boldsymbol{\mu}]\end{aligned} \tag{8.5}$$

where the last equality comes from applying equation (8.4), and the sum over s ranges from x to x + w − n in steps of n years. [For convenience, equation (8.5) assumes that w is an integral multiple of n and that all age intervals have length n, but those restrictions can easily be relaxed.] Now let **y(x)** be a k element column vector whose jth element, $y_j(x)$, gives the number of persons in state j at exact age x. From equations (8.4) and (8.5), we can write the projection equation

$$\mathbf{y}(\mathbf{x}+\mathbf{w}) = \mathbf{P}(\mathbf{x}, \mathbf{w})\ \mathbf{y}(\mathbf{x}) = \exp[\mathrm{n}\Sigma\ \mathrm{z}(\mathrm{s}, \mathrm{n})\boldsymbol{\mu}]\ \mathbf{y}(\mathbf{x}) \tag{8.6}$$

Equation (8.6) is the general PTR solution for advancing a cohort from any age x to any age x + w. Base rate matrix $\boldsymbol{\mu}$ can be of any form as long as it reflects demographically realistic rates (i.e. non-negative, finite, real numbers). As x + w becomes large, **P(x, w)** becomes a rank one matrix and **y(x + w)** approaches the stable state composition implied by the dominant right eigenvector of $\boldsymbol{\mu}$. That does not arise in cohort analysis, however, because convergence to stability typically requires considerably longer than a single lifetime.

It is useful to examine the PTR equations using eigenstructure relationships, i.e. $\boldsymbol{\mu} = \mathbf{URV}$, $\boldsymbol{\Lambda} = \exp[\mathrm{n}\mathbf{R}]$, and $\mathbf{A} = \mathbf{U\Lambda V}$ (cf. Equation (A.11)). With cumulative level factor Z(x) defined by

$$\mathrm{Z}(\mathrm{x}) = \sum_{\mathrm{s}=0}^{\mathrm{x}-1} \mathrm{z}(\mathrm{s}) \tag{8.7}$$

the eigenstructure relationships and equation (8.6) yield

$$\mathbf{y}(\mathbf{x}) = \mathbf{U}\boldsymbol{\Lambda}^{\mathbf{Z}(\mathbf{x})}\mathbf{V}\ \mathbf{y}(\mathbf{0}) = \mathbf{A}^{\mathbf{Z}(\mathbf{x})}\ \mathbf{y}(\mathbf{0}) \tag{8.8}$$

The k element vector of person-years lived by the cohort in each state at age x can be denoted by $\mathbf{Y(x)}$, where jth element $Y_j(x)$ represents the number of person-years lived in state j at age x last birthday. The value of $\mathbf{Y(x)}$ can be found by integrating $\mathbf{y(x)}$ over age from x to $x+1$. With $\mathbf{I}$ the $k \times k$ identity matrix, the eigenstructure relations and equation (8.8) yield

$$\begin{aligned}\mathbf{Y(x)} &= \mathbf{U}[\textstyle\int e^{sz(x)\mathbf{R}}ds]\mathbf{V}\ \mathbf{y(x)} \\ &= \mathbf{U}(\mathbf{\Lambda}^{z(x)} - \mathbf{I})\mathbf{R}^{-1}\mathbf{V}\ \mathbf{y(x)}/z(x) \end{aligned} \qquad (8.9)$$

Paralleling equations (1.10) and (1.62), the elements of $\mathbf{Y}$ can be used to determine the number of decrements between states in an age interval. For example, the number of decrements from state j to state i between the ages of x and $x+1$ is given by the product of $Y_j(x)$ times $z(x)\mu_{ji}$. The $Y_j$ can also be used in calculating population based life expectancies (cf. equation (1.71)).

Using eigenstructure facilitates sensitivity analyses with respect to cumulative level factor Z. The eigenstructure relationships and equation (8.8) lead to

$$d\ \mathbf{y(x)}/d\ Z(x) = n\mathbf{U}\ \mathbf{\Lambda}^{Z(x)}\mathbf{R}\ \mathbf{V}\ \mathbf{y(0)} \qquad (8.10)$$

As in equation (8.9), all roots of the base rate matrix play an important role in the result.

Graphically, a plot showing how state composition $\mathbf{y}$ varies with Z, what Schoen and Canudas-Romo (2006) termed a life course schematic, can have considerable heuristic value. Such a schematic indicates all of the possible state distributions that a cohort can manifest as its members move through life under the pattern of rates specified by $\boldsymbol{\mu}$ (see Figure 8.1).

In the PTR model of equations (8.1) and (8.2), mortality rates are subject to the same proportionality factors as the rates of interstate transfer. An alternative way to incorporate mortality is to assume that there is an equal risk of dying in all states. Survivorship can then be added as an "exogenous" multiplicative factor applied to all states. With death rates excluded from base rate matrix $\boldsymbol{\mu}$, the age-specific rate matrix, including mortality, can be written

$$\mathbf{M(x, n)} = z(x, n)\boldsymbol{\mu} - \mu_\delta(x, n)\mathbf{I}$$

where $\mu_\delta(x, n)$ is the death rate in all states between the ages of x and $x+n$. That modification changes the total number of survivors but not the proportion in each state, and allows great flexibility with regard to recognizing age-specific mortality.

*Figure 8.1. Life Course Schematic of Parity Distributions for Z Values From 0 to 10, United States Females, 1997*

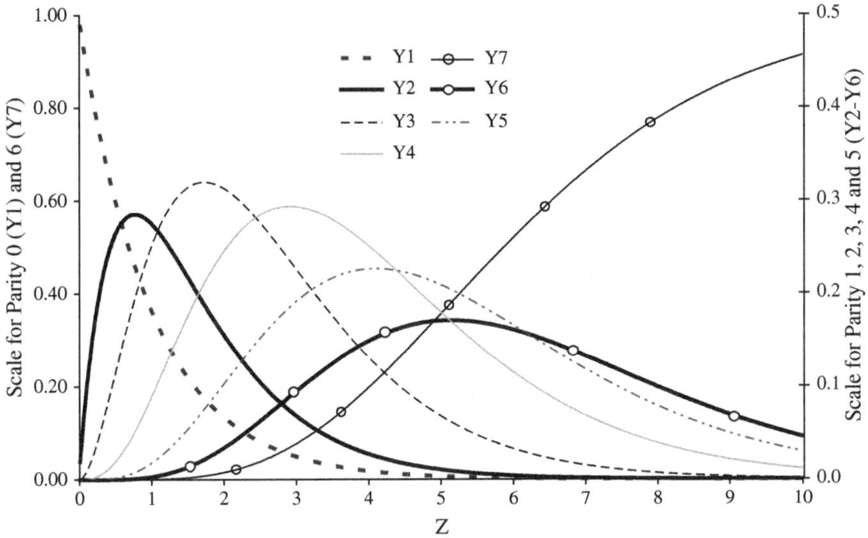


*Source:* Schoen and Canudas-Romo (2006).

### 8.2.2 PTR Models With 2 Living States

Consider the general model with two living states, designated states $\underline{1}$ and $\underline{2}$. The base rate matrix can be written

$$\boldsymbol{\mu} = \begin{bmatrix} -\mu_{12} - \mu_{1\delta} & \mu_{21} \\ \mu_{12} & -\mu_{21} - \mu_{2\delta} \end{bmatrix} \tag{8.11}$$

Persons in the model can move back and forth between states $\underline{1}$ and $\underline{2}$ and from either state to absorbing state $\underline{\delta}$.

The PTR solution can be shown in scalar terms when the eigenstructure of $\boldsymbol{\mu}$ is relatively simple, as is the case here. The eigenvalue matrix of $\boldsymbol{\mu}$ is given by

$$\mathbf{R} = \begin{bmatrix} \mathrm{r}_1 & 0 \\ 0 & \mathrm{r}_2 \end{bmatrix} \tag{8.12}$$

and equation (A.4) gives the roots as

$$\mathrm{r}_{1,2} = {}^1\!/_2[-\mu_{12} - \mu_{1\delta} - \mu_{21} - \mu_{2\delta} \pm \{(\mu_{12} + \mu_{1\delta} - \mu_{21} - \mu_{2\delta})^2 + 4\,\mu_{12}\mu_{21}\}^{1/2}] \tag{8.13}$$

where $r_1$ is associated with the positive square root. The right eigenvector matrix is then

$$\mathbf{U} = \begin{bmatrix} 1 & 1 \\ \dfrac{\mu_{12}}{r_1 + \mu_{21} + \mu_{2\delta}} & \dfrac{\mu_{12}}{r_2 + \mu_{21} + \mu_{2\delta}} \end{bmatrix} \qquad (8.14)$$

(cf. equation (A.5)), with $\mathbf{U}^{-1} = \mathbf{V}$. Given n year age intervals, the eigenvalue matrix of $\mathbf{A}$ is

$$\boldsymbol{\Lambda} = \begin{bmatrix} \exp(n\ r_1) & 0 \\ 0 & \exp(n\ r_2) \end{bmatrix} \qquad (8.15)$$

with diagonal elements $\lambda_j = \exp(n\ r_j)$.

Assume that at age 0 the population consists of one person in state $\underline{1}$ and no persons in state $\underline{2}$. Equation (8.8) then yields the solution

$$\mathbf{y}(\mathbf{x}) = \begin{bmatrix} \{\lambda_1{}^{Z(x)}[r_1 + \mu_{21} + \mu_{2\delta}] - \lambda_2{}^{Z(x)}[r_2 + \mu_{21} + \mu_{2\delta}]\}/(r_1 - r_2) \\ \mu_{12}[\lambda_1{}^{Z(x)} - \lambda_2{}^{Z(x)}]/(r_1 - r_2) \end{bmatrix} \qquad (8.16)$$

Equation (8.16) shows how the age-specific size and state trajectory of the two living state model are driven by Z(x), the sole function of age on the right-hand side of equation (8.8). Both the dominant and subordinate roots of $\boldsymbol{\mu}$ play significant roles in model dynamics.

### 8.2.3 Hierarchical PTR Models

Hierarchical models require interstate movements to follow a given ordering, and do not permit returns to a previously occupied state. They arise frequently in demographic analysis. Examples include multistate models with the marital statuses Never Married and Ever Married; the health statuses Susceptible, Infected, and Recovered/immune; and the parity statuses having 0 children, having 1 child, having 2 children, etc. Hierarchical models may be particularly suitable for PTR analyses because it is often reasonable to assume that all rates move proportionally over age, and in many instances the eigenstructure of the underlying rate matrix is relatively simple.

The base transfer rate matrix of a strictly hierarchical model with any number of living states can be written

$$\boldsymbol{\mu} = \begin{bmatrix} -\mu_{12} & 0 & 0 & 0 & \cdots \\ \mu_{12} & -\mu_{23} & 0 & 0 & \cdots \\ 0 & \mu_{23} & -\mu_{34} & 0 & \cdots \\ 0 & 0 & \mu_{34} & -\mu_{45} & \cdots \end{bmatrix} \qquad (8.17)$$

Persons can move from state $\underline{1}$ to state $\underline{2}$, from state $\underline{2}$ to state $\underline{3}$, from state $\underline{3}$ to state $\underline{4}$, and so on. The model can accommodate any number of states, as long as all moves are from state j to state $j+1$. Mortality can be incorporated most easily as an exogenous survivorship factor that affects all states to the same degree.

The base rate matrix of equation (8.17) is a lower triangular matrix, hence its roots are its diagonal elements (i.e. minus the state specific transfer rates). The right eigenvector matrix is patterned and can be written rather compactly (see Exercise 2). If we again assume that the age 0 population has just one person in state $\underline{1}$, equation (8.8) leads to the state trajectories

$$y_1(x) = \exp[-n\, Z(x)\mu_{12}]$$

$$y_2(x) = \mu_{12}\{\exp[-n\, Z(x)\mu_{23}] - \exp[-n\, Z(x)\mu_{12}]\}/(\mu_{12} - \mu_{23})$$

$$y_3(x) = \frac{[\mu_{12}\mu_{23}]\{(\mu_{12} - \mu_{23})\exp[-nZ(x)\mu_{34}] - (\mu_{12} - \mu_{34})\exp[-nZ(x)\mu_{23}] - (\mu_{23} - \mu_{34})\exp[-nZ(x)\mu_{12}]\}}{(\mu_{12} - \mu_{23})(\mu_{12} - \mu_{34})(\mu_{23} - \mu_{34})}$$

and in general

$$y_j(x) = [\prod_{i=1}^{j-1} \mu_{i,i+1}] \sum_{i=1}^{j} \{\exp[-n\, Z(x)\mu_{i,i+1}] / \prod_{\substack{h=1 \\ h\neq i}}^{j} (\mu_{h,h+1} - \mu_{i,i+1})\} \qquad (8.18)$$

Schoen and Canudas-Romo (2006) employed the solution in equation (8.18) to model parity progression, using a six living state model that recognized the parities 0, 1, 2, 3, 4, and 5 or more. The relatively few births to women at parities 6 and over were ignored, as was mortality under the highest age at childbearing. Using data for women in the United States 1997, they found the proportionality assumption to be approximately true for all parities.

To specify the proportionalized rates, Schoen and Canudas-Romo (2006) first preserved the observed age-specific transfer rates from parity 0 to parity 1. The fixed ratio of the base rate at parity j to the base rate at parity 0 was determined by the ratio of the observed rates at ages 26–30. That interval is close to the mean age at childbearing, and produced proportionalized rates with a TFR almost identical to the TFR of the observed rates. The principal deviations from proportionality occurred at the younger ages, where rates at higher parities (which reflected the behavior of relatively few women) were disproportionately large. Comparing the state trajectories implied by the observed and proportionalized rates, Schoen and Canudas-Romo (2006a) found substantial similarities, the major departure being that the proportionalized rates overstated the age 25 proportion at parity 1 by 0.05. On balance, they considered the proportionalized rates sufficiently similar to the observed rates to justify their illustrative use.

Figure 8.1 shows the life course schematic of the PTR model, i.e. the proportion in each parity state by cumulative level function Z. The single year of age occurrence/exposure rates of first birth in the base matrix were normalized to 1, so the Z function represents the cumulative sum of the rates of transfer to parity 1. The observed TFR in the US 1997 was 2.09, which under the PTR base $\boldsymbol{\mu}$ matrix is attained at a Z value of 1.91. The PTR proportion childless at that Z level is then exp[−1.91] or 0.148 (cf. equation (8.18)). As Z increases, the proportion at parity 0 declines exponentially from 1 to zero, while the proportion in "absorbing" parity 6 increases sigmoidally from 0 to 1 (see scale at the left of Figure 8.1). The proportions at parities 1 through 5 are all unimodal with a skew to the right (scale at the right of Figure 8.1).

Given the level of the 1997 rates, parity 2 has the highest maximum proportion of any parity, 0.320, which is attained when $Z = 1.74$. The maxima for parities 3, 4, and 5 are at successively smaller proportions of the total cohort, and all are attained at Z values above the level reached by the 1997 US rates. Since women of parity 3 comprise the largest fraction of the model cohort for Z values from 2.46 to 4.27 (or TFRs of 2.61 to 4.05), fertility would have to be extraordinarily high by contemporary Western standards before the most common parity would exceed 3. There is no Z value at which women of parity 1 comprise the largest fraction of the cohort. When Z is 1.25 (and the TFR is 1.40) the most common parity shifts from 0 to 2.

It is clear from Figure 8.1 that the slope of a parity trajectory varies markedly with Z. The slope of the parity 0 trajectory is always negative (though it declines as Z increases), the slope of the parity 6 trajectory is always positive (though to differing degrees), and the slopes of the other parity trajectories are positive before their mode and negative thereafter. Assume that all of the US 1997 rates had been a bit higher at some age so as to retain the same $\mu$ but attain a slightly larger Z. Then the ultimate proportion at parities 0–2 would decrease and the ultimate proportion at parities 3 and higher would increase, with the proportion at parity 3 increasing the most.

The effect of changing the relative size of one of the base rates can be found by differentiating equation (8.18) with respect to that base rate. For rates to parities 1 through 5, the pattern of effects is similar and worth examining. Let us look at changes in $\mu_{12}$, the base transfer rate from parity 0 to parity 1, and focus on $y_2$, the proportion at parity 1. From equation (8.18) we obtain

$$\begin{aligned} \mathrm{d}\, y_2(Z)/\mathrm{d}\, \mu_{12} = \{&\exp[-nZ\mu_{12}][\mu_{23} + \mu_{12}(\mu_{12} - \mu_{23})Z] \\ &- \mu_{23} \exp[-n\, Z\mu_{23}]\}/(\mu_{12} - \mu_{23})^2 \end{aligned} \tag{8.19}$$

a weighted difference between exponential terms in $\mu_{12}$ and $\mu_{23}$. Setting the derivative equal to zero and rearranging terms gives

$$Z = \ln\{1 + [Z\ \mu_{12}(\mu_{12} - \mu_{23})/\mu_{23}]\}/[n(\mu_{12} - \mu_{23})] \qquad (8.20)$$

an iterative equation for the extremum of Z. Numerical analyses of equation (8.20) indicate that there is a unique positive value of Z associated with zero change in $y_2(Z)$. At low values of Z, an increase in $\mu_{12}$ raises the proportion at parity 1, but as Z increases that effect is reduced and then reversed, with $y_2(Z)$ ultimately becoming smaller than it was under the original $\mu_{12}$.

Figure 8.2 shows how the proportions at parities 0 through 3 vary from their Figure 8.1 values when $\mu_{12}$ is increased or decreased by 25% of its base rate matrix value. Panel A shows that the proportion at parity 0 varies inversely with $\mu_{12}$. In Panel B, a 25% increase in $\mu_{12}$ yields a larger proportion at parity 1 when

*Figure 8.2. Proportion at Parities 0, 1, 2 and 3 Under Proportionalized Rates for United States Females, 1997 With Alternative Base Rates of Movement From Parity 0 to 1*

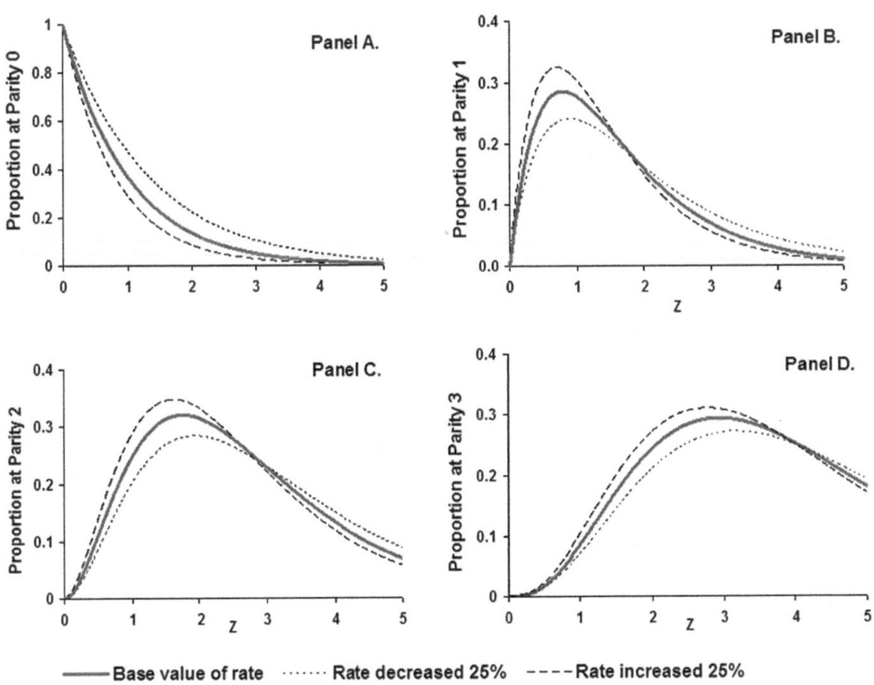


*Source:* Schoen and Canudas-Romo (2006).

Z is less than 1.59, but a smaller proportion for higher values of Z. Panels C and D show qualitatively similar effects on proportions at parities 2 and 3. A larger base $\mu_{12}$ initially produces a higher proportion in the state, but the effect reverses at some higher level of Z.

That pattern of change may seem counterintuitive, but it is rooted in the dynamics of the hierarchical PTR model. It is clear that an increase in $\mu_{12}$ initially increases the proportion at parity 1 because the pace of transfer from parity 0 is accelerated. However, that larger fraction at parity 1 is immediately exposed to the risk of transfer to parity 2. As Z increases, the smaller proportion at parity 0 adds fewer persons to parity 1, while $\mu_{23}$ continues to reduce the increased fraction at parity 1. At some point, the smaller number of entrants to parity 1 and the greater (longer) exposure to the rate of transfer to parity 2 lowers the fraction at parity 1 below its level under the original $\mu_{12}$. The increase in $\mu_{12}$, by accelerating transfers from parity 1, generates the same dynamic at parities 2 and above, though the crossover to lower proportions at those parities than under the initial base value occurs at ever higher levels of Z.

### 8.2.4 A PTR Model of Marriage, Divorce, and Remarriage

Another potential application of PTR models is in the area of marriage, divorce, and remarriage. Marriage and remarriage rates generally move in the same direction, and divorce rates often move in that direction as well.

Schoen and Canudas-Romo (2005b) defined the base rate matrix for a three living state model of first marriage, divorce, and remarriage as

$$\boldsymbol{\mu} = \begin{bmatrix} -\mu_{sm} & 0 & 0 \\ \mu_{sm} & -\mu_{mv} & \mu_{vm} \\ 0 & \mu_{mv} & -\mu_{vm} \end{bmatrix} \qquad (8.21)$$

The three model states are Never Married (s), Presently Married (m), and Divorced (v), and moves can be made from state s to state m (first marriages), from state m to state v (divorces), and from state v to state m (remarriages). To simplify the model, the focus is on ages 15 through 50, with mortality (and widowhood) disregarded.

The model of equation (8.21) is not hierarchical, but the eigenstructure is still fairly simple. The roots of $\boldsymbol{\mu}$ are 0, $-\mu_{sm}$, and $-\mu_{mv} - \mu_{vm}$. The right eigenvector matrix can be written

$$\mathbf{U} = \begin{bmatrix} 0 & 0 & (\mu_{sm} - \mu_{mv} - \mu_{vm})/\mu_{mv} \\ \mu_{vm}/\mu_{mv} & -1 & (\mu_{vm} - \mu_{sm})/\mu_{mv} \\ 1 & 1 & 1 \end{bmatrix} \qquad (8.22)$$

With the initial population vector, **y(15)**, consisting of one person in state s̲, equation (8.8) yields

$$\mathbf{y}(\mathbf{x}) = \begin{bmatrix} \exp[-Z(x)\mu_{sm}] \\ \dfrac{\mu_{vm}}{\mu_{vm}+\mu_{mv}} - \dfrac{\exp[-Z(x)(\mu_{vm}+\mu_{mv})]\mu_{sm}\mu_{mv}}{(\mu_{vm}+\mu_{mv}-\mu_{sm})(\mu_{vm}+\mu_{mv})} - \dfrac{\exp[-Z(x)\mu_{sm}](\mu_{vm}-\mu_{sm})]}{(\mu_{vm}+\mu_{mv}-\mu_{sm})} \\ \dfrac{\mu_{mv}}{\mu_{vm}+\mu_{mv}} + \dfrac{\exp[-Z(x)(\mu_{vm}+\mu_{mv})]\mu_{sm}\mu_{mv}}{(\mu_{vm}+\mu_{mv}-\mu_{sm})(\mu_{vm}+\mu_{mv})} - \dfrac{\exp[-Z(x)\mu_{sm}]\mu_{mv}}{(\mu_{vm}+\mu_{mv}-\mu_{sm})} \end{bmatrix} \tag{8.23}$$

where the first element refers to state s̲, the second to state m̲, and the third to state v̲. In the absence of mortality, the sum of the three elements of **y(x)** is always 1. Each element is a weighted sum involving exponentiated eigenvalues of $\boldsymbol{\mu}$.

Schoen and Canudas-Romo (2005b) applied the PTR model to data for United States women in 1995. They started with the observed single year of age first marriage rates, and determined the relative sizes of the divorce and remarriage rates based on the ratios observed at ages 34–38. Those ages were chosen to provide the closest agreement between the proportionalized (or "stylized") rates and the observed rates over the 15–49 year age range. The base first marriage rates were then scaled to 1, yielding ratios of 1 : 0.4491 : 1.8701 that were used as the elements of the base rate matrix.

Figure 8.3 shows a reasonably good agreement between the observed and stylized rates. The largest departures from proportionality involve rates based on small fractions of the cohort, especially married women under age 20 and divorced women in their teens and early 20s. Schoen and Canudas-Romo (2005b) also found that the observed and stylized state trajectories were quite similar. At age 50, where agreement was poorest, the proportions married differed by only 0.013 and the proportions divorced by only 0.007.

Figure 8.4 provides the life course schematic for the marriage-divorce-remarriage model. With the first marriage rates scaled to 1 in the $\mu$ matrix, Z(x) is the sum of the age-specific first marriage rates up to that age. At age 50, the PTR proportion never married is 0.116, indicating that $0.116 = \exp[-Z(50)]$ or that $Z(50) = 2.154$. As indicated in equation (8.23), the cohort proportion Never Married declines exponentially with Z. At high values of Z, the proportion in state s̲ is essentially zero, and the relative proportions in states m̲ and v̲ approach the stable ratio $\mu_{vm}/\mu_{mv}$ (see equation (8.22)). In the present model that ratio is $1.8701/0.4491 = 4.164$; eventually the proportion in state m̲ becomes 0.806 and the proportion in state v̲ becomes 0.194. Figure 8.4 shows that the proportions in both states grow monotonically from 0 to their long term values. (Under

*Figure 8.3. Observed (obs) and Stylized (sty) Rates of First Marriage, Divorce, and Remarriage, with the Stylized Rates Based on Ages 34 to 38, United States Females, 1995*

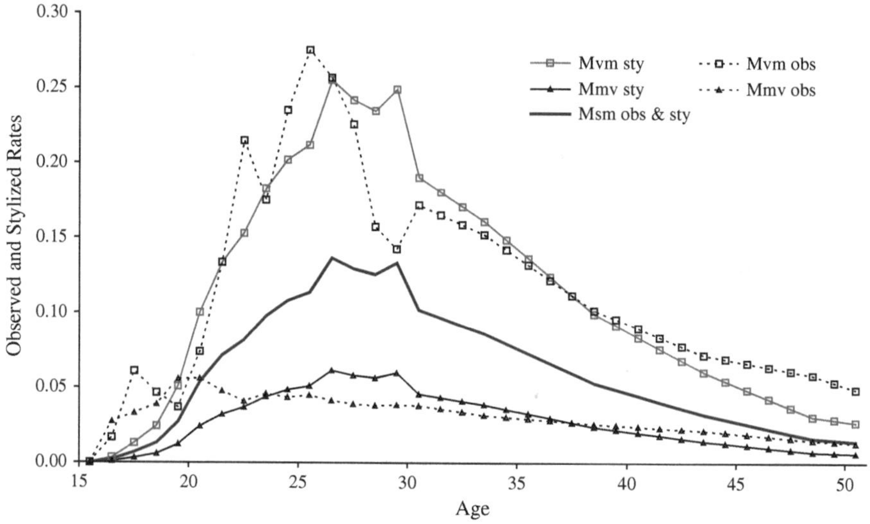


*Source:* Schoen and Canudas-Romo (2005b).

certain conditions, however, the trajectory in the Presently Married state can be nonmonotonic; see Exercise 4.)

Schoen and Canudas-Romo (2005b) examined how the proportions in each state would change in the event of a 25% increase or decrease in the relative size of each base rate. Given the US 1995 values, the base rate of remarriage is 87% higher than the base rate of first marriage. Nonetheless, changes in the base remarriage rate have only modest effects. The base rate of divorce is 45% of the base first marriage rate. Changes in divorce rates have a substantial impact on the proportion divorced, but not on the proportion married. Changes in first marriage rates, however, have a substantial impact on the proportion in every marital status. They concluded that the large proportions presently married and small proportions never married observed in the United States are primarily the result of continuing high rates of first marriage.

## 8.3 MULTISTATE METASTABLE MODELS

To specify metastable models in the multistate context, let us begin with a $k \times k$ base $\boldsymbol{\mu}$ rate matrix of the form of equation (8.1), which can be written

*Figure 8.4. Life Course Schematic Showing Proportions Never Married, Married, and Divorced for Z Values From 0 to 6, United States Females, 1995*

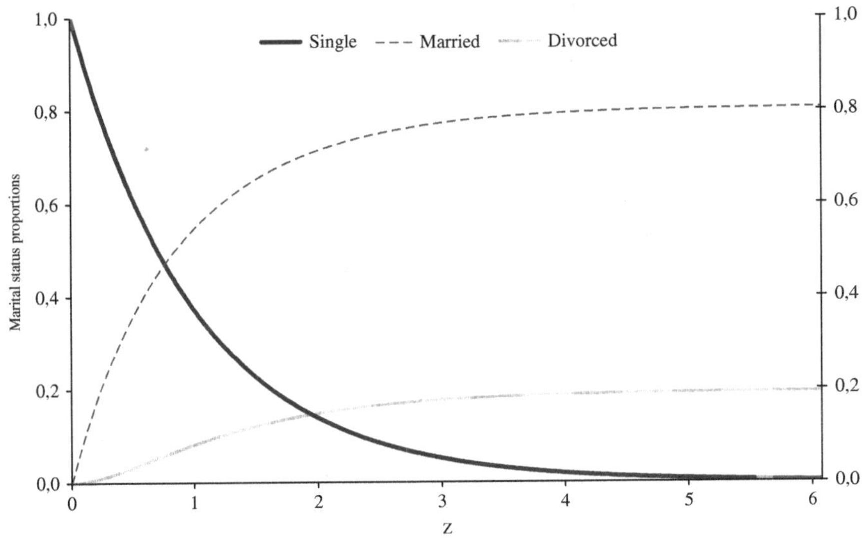


*Source:* Schoen and Canudas-Romo (2005b).

in terms of its eigenstructure as $\boldsymbol{\mu} = \mathbf{URV}$. Given time intervals of n years, the associated survivorship (or projection) matrix is $\mathbf{A} = \mathbf{U\Lambda V}$, where $\boldsymbol{\Lambda} = \exp(n\mathbf{R})$. As in equation (8.4), the (i,j)th element of $\mathbf{A}$, $\mathbf{a}_{ij}$, yields the number of persons in state i at the end of the n year interval per person in state j at the beginning of the interval.

To extend metastability, we introduce k × k non-negative matrix $\mathbf{H}$. Let

$$\mathbf{A}_t = \mathbf{H}^t\mathbf{A}\ \mathbf{H}^{-t} \quad (8.24)$$

It follows that product matrix $\mathbf{P}_{0,t} = \mathbf{A}_t\mathbf{A}_{t-1}\ldots\mathbf{A}_1$ is then given by

$$\mathbf{P}_{0,t} = \mathbf{H}^t(\mathbf{AH}^{-1})^t \quad (8.25)$$

with metastable synthesis matrix $\mathbf{S} = (\mathbf{AH}^{-1)}$.

Although the formal extension is quite straightforward, there are significant substantive differences. The nature of the change in the elements of $\mathbf{A}$ in the birth-death case was closely linked to both the dominant eigenvalue of $\mathbf{A}$ and a demographically realistic pattern of change in net maternity. Neither applies in the multistate case. Assume that $\mathbf{H}$ is a diagonal matrix with jth diagonal element $h_j$. Then at every time point, premultiplying by $\mathbf{H}^t$ and postmultiplying by $\mathbf{H}^{-t}$ leaves the diagonal elements of $\mathbf{A}$ unchanged. However, the (i,j)th off-diagonal element becomes $h_i{}^t h_j{}^{-t}\, a_{ij}$. That increase or decrease in the size of the (i,j)th element impacts both risks of interstate transfer and rates of population growth, because one person in state j at the beginning of the interval now implies a different total number of persons at the end of the interval. Analysts using the model need to verify that such changes are appropriate.

Nothing in the derivation of equation (8.25) prevents $\mathbf{H}$ from having nonzero off-diagonal elements. That opens new analytical possibilities that have not been studied in any detail. Still, it is clear that nonzero off-diagonal elements change some of the mathematical properties of the model. As indicated in Exercise 6, the multistate metastable model with diagonal $\mathbf{H}$ matrices maintains constant age-specific momentum over time, and exhibits a time invariant Kullback distance between the metastable population and its associated stable population. Those properties are not preserved when $\mathbf{H}$ has nonzero off-diagonal elements.

## 8.4 MULTISTATE INTRINSICALLY DYNAMIC MODELS

In multi-age IDM models, requiring constant subordinate eigenvalues is enough to assure the multiplicative property (i.e. that the dominant root of the product matrix equals the product of the individual PPM dominant roots). That requirement is also sufficient to generate PPM elements and to specify an analytically soluble dynamic model. In the multistate context, additional restrictions are needed to obtain an analytical solution, because multi-age models impose strong conditions not present in multistate models. Assuming equal rates of natural increase across all living states implies a rate matrix with constant column sums, and satisfies the multiplicative property. However, uniform natural increase alone is not enough to yield tractable dynamic models. In this section (except for new results in subsection 8.4.2), we draw on Schoen (2003), which provides state trajectories from rates for a restricted set of multistate IDMs.

### 8.4.1 The Restricted Two Living State Uniform Growth Model

Let us start by specifying a continuous two living state model in terms of the differential equations

$$d\ell_1(t)/dt = -m_{12}(t)\ell_1(t) + m_{21}(t)\ell_2(t)$$
$$d\ell_2(t)/dt = -m_{21}(t)\ell_2(t) + m_{12}(t)\ell_1(t) \qquad (8.26)$$

where $\ell_j(t)$ represents the number of persons in state j at time t and $m_{ij}(t)$ represents the instantaneous rate (or force) of transfer from state i to state j at time t. The model of equations (8.26) allows movements between states $\underline{1}$ and $\underline{2}$, but no natural increase in either state (i.e. each state's birth rate minus its death rate is zero). The absence of growth in either state is manifested in the fact that the sum of the two equations in (8.26) is always zero.

Without loss of generality we can scale the model so that

$$\ell_1(t) + \ell_2(t) = 1 \qquad (8.27)$$

Equation (8.27) can be used in equation (8.26) to yield the single differential equation

$$d\ell_1(t)/dt = -[m_{12}(t) + m_{21}(t)]\ell_1(t) + m_{21}(t) \qquad (8.28)$$

Equation (8.28) is an elementary first order differential equation whose solution is (cf. Ford 1955)

$$\ell_1(t) = \ell_1(0)\, e^{-H(t)} + e^{-H(t)} \int_0^t e^{H(s)} m_{21}(s)\, ds \qquad (8.29)$$

where

$$H(t) = \int_0^t [m_{12}(u) + m_{21}(u)]\, du$$

The integration in equation (8.29) is very difficult to perform unless $[m_{12}(u) + m_{21}(u)]$ is constant over time. That sum of rates represents the "metabolism" or rate of turnover in the model; assuming a constant metabolism is often quite reasonable. Imposing that restriction gives the solution

$$\ell_1(t) = \ell_1(0)e^{-Mt} + e^{-Mt} \int_0^t e^{Ms}\, m_{21}(s)\, ds \qquad (8.30)$$

where the fixed turnover rate is given by

$$M = m_{12}(t) + m_{21}(t)$$

If $m_{21}$ is a polynomial, exponential, or trigonometric function, the integral in equation (8.30) can be evaluated analytically.

Let $m_{21}(t)$ be a simple exponential function of time, i.e.

$$m_{21}(t) = m_{21}(0)\, e^{at} \tag{8.31}$$

In the long term, the $\ell_1(0)e^{-Mt}$ term on the right hand side of equation (8.30) goes to zero because M must be a positive number if either rate is nonzero. Substantively, the attenuation of that term describes how the initial state composition of the model is forgotten. In the long term, with parameters that yield non-negative transfer rates, the population composition at time t under equation (8.31) is

$$\ell_1(t) = m_{21}(0)\, e^{at}/(M + a) \tag{8.32}$$

The fraction in state 1 changes exponentially over time. It goes to zero if $a < 0$ and to infinity if $a > 0$, though the model ceases to be valid demographically when $\ell_1(t) > 1$.

Models with cyclical change are of particular interest, and with suitable parameters remain demographically realistic indefinitely. With $|b| < 1$, let us assume that

$$m_{21}(t) = m_{21}(0)[1 + b\sin(\omega t)] \tag{8.33}$$

In the long term when $e^{-Mt}$ goes to zero, equation (8.30) yields

$$\ell_1(t) = m_{21}(0)\{(1/M) + b[M\sin(\omega t) - \omega\cos(\omega t)]/[M^2 + \omega^2]\} \tag{8.34}$$

Equation (8.34) shows that the proportion in state $\underline{1}$ cycles around its stable fraction, i.e. around $m_{21}(0)/M$. The fluctuations have cycle length $T = 2\pi/\omega$, and are determined by a weighted linear combination of $\sin(\omega t)$ and $\cos(\omega t)$.

Figure 8.5 shows how the fraction in state $\underline{1}$ and the normalized transfer rate $m_{21}(t)/M$ vary over time in a typical sinusoidal model. From equation (8.33), the first maximum of $m_{21}(t)$ after $t = 0$ comes at time T/4. The $\ell_1(t)$ curve always leads the $m_{21}(t)/M$ curve, though the sinusoidal fluctuations in the latter always have a greater amplitude. Both curves are symmetric and have the same cycle length. They intersect twice during each cycle, at the maximum and minimum points of the $\ell_1(t)$ curve. At those points, the dynamic model and its associated stable population have identical rates and state compositions (see Exercises 7 and 8).

To this point, we have only considered the model of equations (8.30) and (8.33) in the long term. The complete solution is

$$\begin{aligned}\ell_1(s+t) = {}&\ell_1(s)e^{-Mt} + m_{21}(0)\{[1 - e^{-Mt}]/M + [b/(M^2 + \omega^2)]\cdot \\ &[M\sin\omega(s+t) - \omega\cos\omega(s+t) - e^{-Mt}(M\sin(\omega s) - \omega\cos(\omega s))]\}\end{aligned} \tag{8.35}$$

*Figure 8.5. Time Trajectories of the Normalized Rate of Transfer from State 2 to State 1 ($m_{21}/M$) and the Fraction of Persons in State 1 ($\ell_1$), Sinusoidal Two Living State, No Growth Intrinsically Dynamic Model*

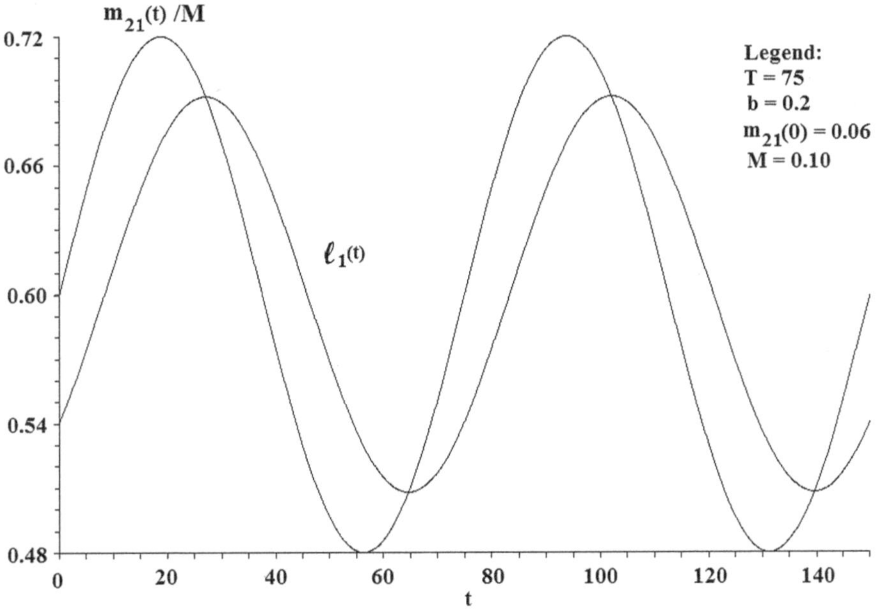


*Source:* Schoen (2003, Figure 1). Reprinted with permission from *Mathematical Population Studies*, Taylor and Francis Group, LLC, http://taylorand francis.com, (c) 2003.

Assume that $\ell_1(s) = 1$, or that the total population of the model is in state $\underline{1}$ at time s. In that case, the value of $\ell_1(s+t)$ given by equation (8.35) is the probability that a person in state $\underline{1}$ at time s will be in state $\underline{1}$ at time $s+t$. We can denote that probability by $p_{11}(s, t)$. If we assume that $\ell_1(s) = 0$, then the value of $\ell_1(s+t)$ given by equation (8.35) is $p_{21}(s, t)$, the probability that a person in state $\underline{2}$ at time s will be in state $\underline{1}$ at time $s+t$. With no births and deaths, it follows that $p_{12}(s, t) = 1 - p_{11}(s, t)$, and $p_{22}(s, t) = 1 - p_{21}(s, t)$, completing the array of probabilities.

The model of equations (8.30) and (8.33) can easily be extended to allow time varying growth, as long as that growth is always the same in all states. Let the growth rate of each state at time t be $\rho(t)$. To introduce growth, add the term $\ell_1(t)\rho(t)$ to the right hand side of the first equation in (8.26) and the term $\ell_2(t)\rho(t)$ to the right hand side of the second equation. The solution to modified equations (8.26) is identical to equation (8.35), except that the entire expression on the right hand side of equation (8.35) is multiplied by $\exp[\int \rho(u)du]$, with the integral ranging from s to s + t. Thus the total size of the model population changes over time, but the state composition remains the same as in the no growth case.

### 8.4.2 Additional Two Living State Models

This subsection extends the preceding discussion in two ways. First, it provides a solution for the discrete version of the no growth constant metabolism model of equation (8.30) for any sequence of discrete transfer rates $m_{21t}$. Second, it derives the discrete state trajectory of a hierarchical two living state model without assuming constant metabolism. An explicit solution for the two living state no growth case, without constant metabolism, can be obtained by straightforward algebra, but is quite complicated. The limiting factor in seeking explicit IDM solutions is the rapidly increasing algebraic complexity of multistate survivorship/projection expressions that span multiple age intervals.

In the discrete case with intervals of one year, model rates at time t can be arrayed in the constant metabolism, no growth rate matrix

$$\boldsymbol{\mu}_t = \begin{bmatrix} -[M - m_{21t}] & m_{21t} \\ M - m_{21t} & -m_{21t} \end{bmatrix} \tag{8.36}$$

Each column of the $\mu$ matrix sums to zero. The dominant and subordinate roots of $\mu_t$, 0 and $-M$, are constant over time. The right eigenvector matrix of $\boldsymbol{\mu}_t$ can be written

$$\mathbf{U}_t = \begin{bmatrix} 1 & 1 \\ \dfrac{M - m_{21t}}{m_{21t}} & -1 \end{bmatrix} \tag{8.37}$$

where the subordinate right eigenvector is constant over time. With $\mathbf{V} = \mathbf{U}^{-1}$, the time t PPM is given by $\mathbf{A}_t = \mathbf{U}_t\ \mathbf{exp(R)}\ \mathbf{V}_t$, or

$$\mathbf{A}_t = \begin{bmatrix} \{m_{21t} + \exp[-M](M - m_{21t})\}/M & m_{21t}(1 - \exp[-M])/M \\ (M - m_{21t})(1 - \exp[-M])/M & \{M - m_{21t}(1 - \exp[-M])\}/M \end{bmatrix} \tag{8.38}$$

Each column of $\mathbf{A}_t$ sums to one.

Finding the product matrix here is reminiscent of the situation encountered in the discrete birth-death model of subsection 7.4.1, and the present derivation is achieved in a similar fashion. The multistate analog of the time t birth-death product matrix mean age at childbearing, $\mu_{Mt}$, is given by the series

$$\begin{aligned} 1/\mu_{Mt} = {} & \{(1 - \exp[-M])/M\}\{m_{21t} + \exp[-M]m_{21,t-1} + \exp[-2M]m_{21,t-2} + \dots \\ & + \exp[-M(t-2)]m_{212} + \exp[-M(t-1)]m_{211}\} \end{aligned} \tag{8.39}$$

Since $M > 0$, earlier terms are increasingly discounted, and $1/\mu_{Mt}$ can be interpreted as a weighted function of $m_{12t}/m_{21t}$ ratios that emphasizes recent behavior. The right eigenvector matrix of product matrix $\mathbf{P}_{0,t} = \mathbf{A}_t\mathbf{A}_{t-1}\dots\mathbf{A}_1$ is then given by

$$\mathbf{U}_{Pt} = \begin{bmatrix} 1 & 1 \\ \mu_{Mt} - 1 & -1 \end{bmatrix} \tag{8.40}$$

The IDM product matrix dominant growth rate is always 1. Its dominant left eigenvector is $\mathbf{v'}_{Mt} = [1\ \ 1]/\mu_{Mt}$, which includes the time varying $\mu_{Mt}$. When the time t population given by the first column of $\mathbf{U}_{Pt}$ is premultiplied by the $\mathbf{A}_{t+1}$ matrix that follows from equation (8.38), consistent time $t+1$ values are obtained, confirming the projection relationship.

Turning to the hierarchical two living state model, its rate matrix can be written

$$\boldsymbol{\mu}_t = \begin{bmatrix} -w_t & 0 \\ w_t & -z_t \end{bmatrix} \tag{8.41}$$

where $w_t$ is the time t rate of transfer from state $\underline{1}$ to state $\underline{2}$, and $z_t$ is the time t rate of movement from state $\underline{2}$ to the absorbing state (and thus out of the model population). By the usual constant forces within age intervals assumption, the survivorship/projection matrix is given by

$$\mathbf{A}_t = \begin{bmatrix} \exp[-w_t] & 0 \\ G_t & \exp[-z_t] \end{bmatrix} \tag{8.42}$$

where

$$G_t = w_t\{\exp[-z_t] - \exp[-w_t]\}/(w_t - z_t)$$

Product matrix $\mathbf{P}_{0,t} = \mathbf{A}_t\ \mathbf{A}_{t-1}\dots\mathbf{A}_1$ can be found by straightforward algebra. The result is

$$\mathbf{P}_{0,t} = \begin{bmatrix} \exp[-\Sigma\ w_j] & 0 \\ H_{0,t} & \exp[-\Sigma\ z_j] \end{bmatrix} \tag{8.43}$$

where the sums over j range from 1 to t and $H_{0,t}$ is given by

$$H_{0,t} = \sum_{j=1}^{t} G_j \exp[-\sum_{i=1}^{j-1} w_i]\ \exp[-\sum_{i=j+1}^{t} z_i]$$

### 8.4.3 The Restricted No Growth IDM With Three Living States

Returning to the Schoen (2003) approach used in subsection 8.4.1, we can specify a three living state, no growth model using the differential equations

$$\begin{aligned} d\ell_1(t)/dt &= -[m_{12}(t)+m_{13}(t)]\ell_1(t)+m_{21}(t)\ell_2(t)+m_{31}(t)\ell_3(t) \\ d\ell_2(t)/dt &= -[m_{21}(t)+m_{23}(t)]\ell_2(t)+m_{12}(t)\ell_1(t)+m_{32}(t)\ell_3(t) \\ d\ell_3(t)/dt &= -[m_{31}(t)+m_{32}(t)]\ell_3(t)+m_{13}(t)\ell_1(t)+m_{23}(t)\ell_2(t) \end{aligned} \tag{8.44}$$

In the absence of growth the model population can be scaled without loss of generality, hence we assume

$$\ell_1(t)+\ell_2(t)+\ell_3(t)=1 \tag{8.45}$$

By substituting equation (8.45) in equation (8.44) to eliminate $\ell_3(t)$ we can write

$$\begin{aligned} d\ell_1(t)/dt &= -[m_{12}(t)+m_{13}(t)+m_{31}(t)]\ell_1(t)+[m_{21}(t)-m_{31}(t)]\ell_2(t)+m_{31}(t) \\ d\ell_2(t)/dt &= -[m_{21}(t)+m_{23}(t)+m_{32}(t)]\ell_2(t)+[m_{12}(t)-m_{32}(t)]\ell_1(t)+m_{32}(t) \end{aligned} \tag{8.46}$$

To simplify the solution, we require the four bracketed sets of rates in equation (8.46) to be constant over time and write

$$\begin{aligned} M_1 &= m_{12}(t)+m_{13}(t)+m_{31}(t) \\ M_2 &= m_{21}(t)+m_{23}(t)+m_{32}(t) \\ M_3 &= m_{21}(t)-m_{31}(t) \\ M_4 &= m_{12}(t)-m_{32}(t) \end{aligned} \tag{8.47}$$

Given equations (8.47), equations (8.46) lead to a second order differential equation where rates $m_{31}$ and $m_{32}$ are free to vary with time. When those rates are simple polynomial, exponential, or trigonometric functions, the differential equations can be solved by established procedures such as the Method of Undetermined Coefficients (cf. Boyce and DiPrima 1977, Ch. 5). In addition to preserving the IDM multiplicative property, equations (8.46) and (8.47) are associated with models whose PPMs have constant subordinate right eigenvectors.

The simplest case occurs when $M_3 = M_4 = 0$. In that "diagonal" case, the $\ell_2(t)$ term falls out of the equation for $\ell_1(t)$, and the $\ell_1(t)$ term falls out of the equation for $\ell_2(t)$. That simplification occurs because the constrained rates of movement into states $\underline{1}$ and $\underline{2}$ do not depend on the state of origin. Consequently, each equation in (8.46) is of the form of equation (8.30) and can be solved independently of the other.

### 8.4.4 The Restricted Uniform Growth IDM With n Living States

The no growth model with n living states can be specified using differential equations of the form

$$d\ell_j(t)/dt = -[\Sigma_i m_{ji}(t)]\ell_j(t) + \Sigma_i\ [m_{ij}(t)\ell_i(t)] \tag{8.48}$$

where the sums range over all living states $i \neq j$. With the total model population equal to 1, $\ell_n(t)$ can be eliminated from equations (8.48), leaving $(n-1)$ equations of the form

$$d\ell_j(t)/dt = -[m_{nj}(t) + \sum_{i=1}^{n} m_{ji}(t)]\ell_j(t) + \sum_{i=1}^{n-1}[m_{ij}(t) - m_{nj}(t)]\ell_i(t) + m_{nj}(t) \tag{8.49}$$

where the sums exclude $i = j$. With n living states, $(n-1)$ rates, i.e. the $m_{nj}(t)$, can be chosen by the analyst. Population growth that varies over time but is uniform across all n states can also be introduced, as discussed at the end of subsection 8.4.1. The remaining $(n-1)^2$ rates are constrained by the $(n-1)^2$ values of constants $M_j$. There is one $M_j$ for each bracketed expression in equation (8.49).

When the $m_{nj}$ are simple polynomial, exponential, or trigonometric functions, equations (8.49) can be solved in closed form by standard techniques for dealing with systems of linear differential equations (e.g. Ford 1955, ch. 4). In general, for the multistate IDMs considered here, polynomial rates yield polynomial state trajectories, exponential rates yield exponential trajectories, and sine or cosine rate functions yield trajectories that are sine and cosine functions.

## 8.5 MULTISTATE HYPERSTABILITY

The hyperstable approach of subsection 7.5.1 has its analog in the multistate case. In this section we derive the multistate hyperstable equations from the basic differential equation of population change, discuss the relationship between the multi-age and multistate models, and further examine the model with two living states. The presentation rests on the analyses in Schoen and Kim (2000) and Schoen (2001).

### 8.5.1 Deriving the Multistate Hyperstable Equations

The basic differential equation underlying a multistate model with n living states can be written

$$\mathbf{x}'(\mathbf{t}) = \boldsymbol{\mu}(\mathbf{t})\ \mathbf{x}(\mathbf{t}) \tag{8.50}$$

where $\mathbf{x(t)}$ is an n element column vector whose jth element, $x_j(t)$ is the number of persons in state j at time t, the prime (′) indicates differentiation with respect to time, and $\boldsymbol{\mu}(\mathbf{t})$ is the $n \times n$ rate matrix of equation (8.1) with the elements specific to time t. Given $\boldsymbol{\mu}(\mathbf{t})$, the solution to equation (8.50) can be expressed in terms of a set of linearly independent vectors $\mathbf{x}^{(1)}(\mathbf{t}), \ldots, \mathbf{x}^{(n)}(\mathbf{t})$ termed a fundamental set of solutions (cf. Boyce and DiPrima, 1977, Ch 7). Choosing such a fundamental set of vectors, we can combine them in $n \times n$ fundamental or solution matrix $\mathbf{X(t)}$, which transforms equation (8.50) into

$$\mathbf{X}'(\mathbf{t}) = \boldsymbol{\mu}(\mathbf{t})\ \mathbf{X}(\mathbf{t}) \qquad (8.51)$$

Paralleling the case when $\boldsymbol{\mu}$ is constant, let us write the solution matrix as

$$\mathbf{X}(\mathbf{t}) = \mathbf{W}(\mathbf{t})\ \mathbf{L}(\mathbf{t}) \qquad (8.52)$$

where $\mathbf{W(t)}$ is an $n \times n$ matrix whose first row elements are 1 and whose (i,j)th element, $i > 1$, is $w_{ij}(t)$, and $\mathbf{L(t)}$ is an $n \times n$ diagonal matrix whose jth element is $l_j(t)$. Matrix $\mathbf{W}$ can be seen as analogous to right eigenvector matrix $\mathbf{U}$, and $\mathbf{L}$ can be seen as analogous to PPM eigenvalue matrix $\boldsymbol{\Lambda}$. Using equation (8.52) in equation (8.51) and differentiating leads to

$$\boldsymbol{\mu}(\mathbf{t}) = \mathbf{W}(\mathbf{t})\check{\mathbf{R}}(\mathbf{t})\ \mathbf{W}^{-1}(\mathbf{t}) + \mathbf{W}'(\mathbf{t})\ \mathbf{W}^{-1}(\mathbf{t}) \qquad (8.53)$$

where $\check{\mathbf{R}}(\mathbf{t})$, which can be seen as analogous to the eigenvalue matrix of $\boldsymbol{\mu}$, is a diagonal matrix defined by

$$\check{\mathbf{R}}(\mathbf{t}) = \mathbf{L}'(\mathbf{t})\ \mathbf{L}^{-1}(\mathbf{t}) \qquad (8.54)$$

The jth diagonal element of $\check{\mathbf{R}}(\mathbf{t})$ is denoted by $\check{r}_j(t)$.

To specify the population projection matrix of this general model, we let $\mathbf{X}_t$ be the discrete counterpart of $\mathbf{X(t)}$ (i.e. the corresponding discrete solution) and write the projection equation

$$\mathbf{X}_{t+1} = \mathbf{A}_{t+1}\mathbf{X}_t \qquad (8.55)$$

where PPM $\mathbf{A}_{t+1}$ is defined analogously to PPM $\mathbf{A}$ in equation (8.24). Let the product matrix $\mathbf{P}_{t,s}$ that projects the population from time t to time $t+s$ be defined by

$$\mathbf{P}_{t,s} = \mathbf{A}_{t+s}\ \mathbf{A}_{t+s-1} \cdots \mathbf{A}_{t+1} \qquad (8.56)$$

Using equations (8.51), (8.55), and (8.56), we have

$$\mathbf{P}_{t,s} = \mathbf{X}_{t+s}\ \mathbf{X}_t^{-1} = \mathbf{W}_{t+s}(\mathbf{L}_{t+s}\mathbf{L}_t^{-1})\mathbf{W}_t^{-1} \qquad (8.57)$$

The product of the PPMs telescopes, leaving only the **W** matrices for times t and t + s and the exponentiated sum of the $\check{\mathbf{R}}$ matrices over that s year time interval. Making the reasonable assumption that the sequence of primitive matrices **A** constitute an ergodic set, initial conditions are ultimately forgotten (Cohen, 1982). Hence in the long term $\mathbf{P}_{t,s}$ is a rank one matrix and the state composition of the model is proportional to the first column of $\mathbf{W}_{t+s}$.

Equation (8.53) is not a new relationship, as it dates back at least to Gantmacher (1959, Vol II, Ch XIV). However, it is an important relationship as it affords a different perspective on multistate dynamics, and suggests defining a matrix **v(t)** by

$$\boldsymbol{\nu}(\mathbf{t}) = \mathbf{W}(\mathbf{t})\check{\mathbf{R}}(\mathbf{t})\ \mathbf{W}^{-1}(\mathbf{t}) \tag{8.58}$$

Matrix $\boldsymbol{\nu}(\mathbf{t})$ can be termed a "latent rate" matrix, as it would become the rate matrix if $\mathbf{W}'(\mathbf{t})$ were zero. In other words, the time t latent rates imply a multistate stable population identical to the model population at time t. From knowledge of $\boldsymbol{\nu}(\mathbf{t})$ one can use equation (8.53) to generate the sequence of $\boldsymbol{\mu}(\mathbf{t})$. However, a demographically reasonable sequence of latent rate matrices need not yield a demographically reasonable sequence of rate matrices.

Unfortunately, from a given sequence of $\boldsymbol{\mu}(\mathbf{t})$, it is generally not possible to find the $\boldsymbol{\nu}(\mathbf{t})$ or to solve differential equation (8.53). An additional constraining assumption is needed, and a number of alternatives are possible. Schoen (2001) considered the Additive Eigenvector assumption, which is briefly described in Exercise 11. Here we focus on the Proportional Eigenvector assumption, which can be written

$$\mathbf{W}(\mathbf{t}) = \mathbf{H}(\mathbf{t})\ \mathbf{C} \tag{8.59}$$

where **H(t)** is an n × n diagonal matrix with first diagonal element $h_1(t) = 1$ and jth diagonal element $h_j(t)$, and **C** is an n × n time invariant matrix whose first row and first column elements are equal to 1 and whose (i,j)th element, i, j, > 1, is $c_{ij}$. The name attached to the assumption reflects the fact that all of the elements in row i of latent rate eigenvector matrix **W(t)** are multiplied by $h_i(t)$, so that at every time the ratio of the (i,j)th element of **W** to the (i,1)st element is $c_{ij}$.

Equation (8.59), along with a known $\boldsymbol{\mu}(\mathbf{t})$ and investigator determined trajectories for 2n − 1 rates, can provide a solution to equation (8.53). Substituting the former equation into the latter yields

$$\boldsymbol{\mu}(\mathbf{t}) = \mathbf{H}(\mathbf{t})\ \mathbf{C}\ \check{\mathbf{R}}(\mathbf{t})\ \mathbf{C}^{-1}\mathbf{H}^{-1}(\mathbf{t}) + \mathbf{H}'(\mathbf{t})\ \mathbf{H}^{-1}(\mathbf{t}) \tag{8.60}$$

The derivatives are therefore confined to (n − 1) diagonal elements of $\boldsymbol{\mu}(\mathbf{t})$. Each of the $n^2$ known elements of $\boldsymbol{\mu}(\mathbf{t})$ provides one scalar equation and, disregarding the derivatives for a moment, there are $n^2$ unknowns: (n − 1) unknown values of $h_j(t)$, n unknown values of $\check{r}_j(t)$, and $(n-1)^2$ unknown values of $c_{ij}$. Using $2n-1$ scalar equations without derivatives, the $h_j(t)$ and $\check{r}_j(t)$ can be expressed in terms of known values of $\boldsymbol{\mu}(\mathbf{t})$ and the constants $c_{ij}$. Since the only unknown values in the equations for the $h_j(t)$ are constants, the $h_j(t)$ can be differentiated in terms of the chosen rate trajectories. The remaining $(n-1)^2$ equations are then available to determine the constants $c_{ij}$. For times other than the initial t, the values of the constants $c_{ij}$ constrain $(n-1)^2$ elements of $\boldsymbol{\mu}$. In short, given an initial rate matrix, an investigator can choose trajectories for $2n-1$ of the $n^2$ rates and from them determine the size and state composition of the dynamic model at every point in time. The full rate matrices are specified as part of the solution, and the investigator needs to be satisfied that the inferred trajectories of the other $(n-1)^2$ rates are demographically appropriate.

### 8.5.2 Relationships Between the Multi-Age and Multistate Hyperstable Models

To relate multi-age and multistate hyperstable models, note that equations (8.55) to (8.57) and (8.59) allow the Proportional Eigenvector PPM to be written

$$\mathbf{A}_t = \mathbf{H}_t\mathbf{C}\,(\mathbf{L}_t\,\mathbf{L}_{t-1}{}^{-1})\mathbf{C}^{-1}\,\mathbf{H}_{t-1}{}^{-1} \qquad (8.61)$$

The analogous multi-age hyperstable PPM, given in equation (7.94), is

$$\mathbf{A}_t = \mathbf{G}_t\mathbf{F}\,\mathbf{G}_{t-1}{}^{-1} \qquad (8.62)$$

Multistate diagonal adjustment matrix **H** parallels multi-age diagonal birth matrix **G**. With constant growth, the matrix product $\mathbf{C}(\mathbf{L}_t\,\mathbf{L}_{t-1}{}^{-1})\mathbf{C}^{-1}$ corresponds to constant Leslie form matrix **F**. The multi-age hyperstable model is thus analogous to the multistate, with the added assumption of proportional eigen*values*. In the multistate context, the proportional eigenvalue assumption adds $(n-1)$ constraints, leaving n rates to the investigator's discretion.

The multi-age ("fixed f") models maintain a constant proportional distribution of births by age of mother. Assuming proportional eigenvalues, multistate hyperstable models maintain a constant proportional distribution of persons by origin state (the analog to mother's age at birth). Thus for every state j, there is a constant proportion of persons in that state at time t who were in any given state i at time $t-1$. The constant proportions feature is more of a restriction in multistate models than in multi-age models, because state composition may be more variable in the multistate case and there is no underlying biological constraint. From a different perspective, hyperstable models with proportional eigenvalues impose a "destination based" criterion, in that movers from every other living state constitute a fixed proportion of each destination state. While often unrealistic, that sort of behavioral model is quite acceptable in some contexts (e.g. birth-death models) and may be useful in others (e.g. analyses of migration). In the absence of the proportional eigenvalue assumption, multistate hyperstable models do not have the constant proportions restriction.

### 8.5.3 A Further Examination of the Dynamic Two Living State Model

Multistate differential equation (8.50) can be solved, using equations (8.53) and (8.58), when the latent rates **v(t)** are known in functional form. Matrices **W(t)** and **L(t)** can be found from the eigenstructure of **v(t)**, and **X(t)** can then be obtained from equation (8.52). Assuming that **W(t)** and **L(t)** are differentiable, $\boldsymbol{\mu}(\mathbf{t})$ can be found from equations (8.53) and (8.54).

Rate matrix $\boldsymbol{\mu}(\mathbf{t})$ can be written in terms of latent rates **v(t)** in the two living state case. If

$$\boldsymbol{\nu}(\mathbf{t}) = \begin{bmatrix} a(t) & b(t) \\ c(t) & d(t) \end{bmatrix} \tag{8.63}$$

and

$$\boldsymbol{\mu}(\mathbf{t}) = \begin{bmatrix} a^*(t) & b^*(t) \\ c^*(t) & d^*(t) \end{bmatrix} \tag{8.64}$$

then

$$\begin{aligned} a^* &= a \\ b^* &= b \\ c^* &= c[1 + \{(a-d)/[(a-d)^2 + 4bc]\}\{D\ \ln[bc/(a-d)^2]\}] \\ d^* &= d + \{(a-d)^2\ D\ \ln[(a-d)/b] + 2bc\ D\ \ln[c/b]\}/[(a-d)^2 + 4bc] \end{aligned} \tag{8.65}$$

where the time index in equation (8.65) is suppressed to simplify the notation and D ln indicates the time derivative of the natural logarithm of the immediately following bracketed function.

The first row elements of $\boldsymbol{\mu}(\mathbf{t})$ are the same as those of **v(t)**. The second row elements of $\boldsymbol{\mu}(\mathbf{t})$ involve derivatives of latent rates, and differ from the corresponding elements of **v(t)** when those derivatives are not equal to zero. If b(t) and c(t) are constant and the difference [a(t) − d(t)] is fixed, $\mathbf{v}(\mathbf{t}) = \boldsymbol{\mu}(\mathbf{t})$.

The latent and model rates are also equal if the ratios $b(t)/[a(t)-d(t)]$ and $c(t)/[a(t)-d(t)]$ are time invariant. When $\mathbf{v(t)} = \boldsymbol{\mu}\mathbf{(t)}$ the derivative of $\mathbf{W(t)}$ is zero, and we have the Proportional Transfer Rate model discussed in section 8.2.

In the Proportional Eigenvector hyperstable model, equations (8.65) simplify because $\ln[bc/(a-d)^2]$ is constant, hence $c^*(t) = c(t)$. Let $\boldsymbol{\mu}\mathbf{(t)}$ be given by

$$\boldsymbol{\mu}\mathbf{(t)} = \begin{bmatrix} \rho_1(t) - m_{12}(t) & m_{21}(t) \\ m_{12}(t) & \rho_2(t) - m_{21}(t) \end{bmatrix} \tag{8.66}$$

where $\rho_j(t)$ is the time t rate of natural increase in state j and $m_{ij}(t)$ is the time t transfer rate from state i to state j. With the hyperstable functions as defined in equations (8.53), (8.54) and (8.59), we can write

$$h_2(t) = \{m_{12}(t)/[(-c_{22})m_{21}(t)]\}^{1/2}$$

$$\check{r}_1(t) = \rho_1(t) - m_{12}(t) + \{m_{12}(t)m_{21}(t)/(-c_{22})\}^{1/2}$$

$$\check{r}_2(t) = \rho_1(t) - m_{12}(t) - \{m_{12}(t)\, m_{21}(t)/(-c_{22})\}^{1/2}$$

$$\rho_2(t) - m_{21}(t) = \rho_1(t) - m_{12}(t) + (1 + c_{22})\{m_{12}(t)\, m_{21}(t)/(-c_{22})\}^{1/2} + D \ln h_2(t) \tag{8.67}$$

With $\rho_1(t)$, $m_{12}(t)$, and $m_{21}(t)$ known functions, the last equation in (8.67) uses the initial conditions value of $\rho_2$ to find $c_{22}$ (which is always negative). When $c_{22}$ is known, that equation is used to obtain $\rho_2(t)$. Even before the value of $c_{22}$ is ascertained, the derivative of $h_2(t)$ can be found using the first equation in (8.67).

To illustrate a dynamic two living state hyperstable model, let

$$a(t) = a^*(t) = a$$

$$b(t) = b^*(t) = b \exp[f \sin(\omega t)]$$

$$c(t) = c^*(t) = 0$$

$$d(t) = d \tag{8.68}$$

If follows from either equations (8.65) or (8.67) that

$$d^*(t) = d - f\omega \cos(\omega t) \tag{8.69}$$

completing the specification of $\boldsymbol{\mu}\mathbf{(t)}$. For a demographically realistic model with two living states, $(d - f\omega) > a$. The known elements of $\mathbf{v(t)}$ yield its right eigenvector matrix

$$\mathbf{W(t)} = \begin{bmatrix} 1 & 1 \\ (d-a)e^{-f\sin(\omega t)}/b & 0 \end{bmatrix} \tag{8.70}$$

The eigenvalue matrix of $\boldsymbol{\nu}(\mathbf{t})$, $\check{\mathbf{R}}(\mathbf{t})$, is constant, with (1,1) element d and (2,2) element a. Using equations (8.54) and (8.57), the long term product matrix from time 0 to time T can be written as

$$\mathbf{P}_{\mathbf{0},\mathbf{T}} = \mathbf{e}^{\mathrm{dT}} \begin{bmatrix} 1 \\ (\mathrm{d}-\mathrm{a})\mathrm{e}^{-\mathrm{f}\sin(\omega \mathrm{T})}/\mathrm{b} \end{bmatrix} [0 \quad \mathrm{b}/(\mathrm{d}-\mathrm{a})] \tag{8.71}$$

Assume the time 0 population, $\mathbf{x_0}$, has 1 person in state 1 and $(\mathrm{d}-\mathrm{a})/\mathrm{b}$ persons in state 2, i.e. that $\mathbf{x_0}$ is a column vector whose elements are those of the first column of $\mathbf{W(0)}$. Then the size and structure of the multistate model at time T are given by

$$\mathbf{x_0} = \mathrm{e}^{\mathrm{dT}} \begin{bmatrix} 1 \\ (\mathrm{d}-\mathrm{a})\mathrm{e}^{-\mathrm{f}\sin(\omega \mathrm{T})}/\mathrm{b} \end{bmatrix} \tag{8.72}$$

Figure 8.6 shows the relative state composition (i.e. number of persons in state 2 per person in state 1) and the total population size for the model of equations (8.68), with parameters $\mathrm{a} = -0.015$, $\mathrm{b} = 0.16$, $\mathrm{d} = 0.005$, $\omega = 0.1$, and $\mathrm{f} = 0.15$. Since the rates of natural increase in both states are constant while $\mathrm{b(t)} = \mathrm{b^*(t)}$ varies sinusoidally, total population size oscillates around an

*Figure 8.6. Relative State Composition ($w_1$) and Total Population Size (N) for a Hierarchical Two Living State Proportional Eigenvector Model*

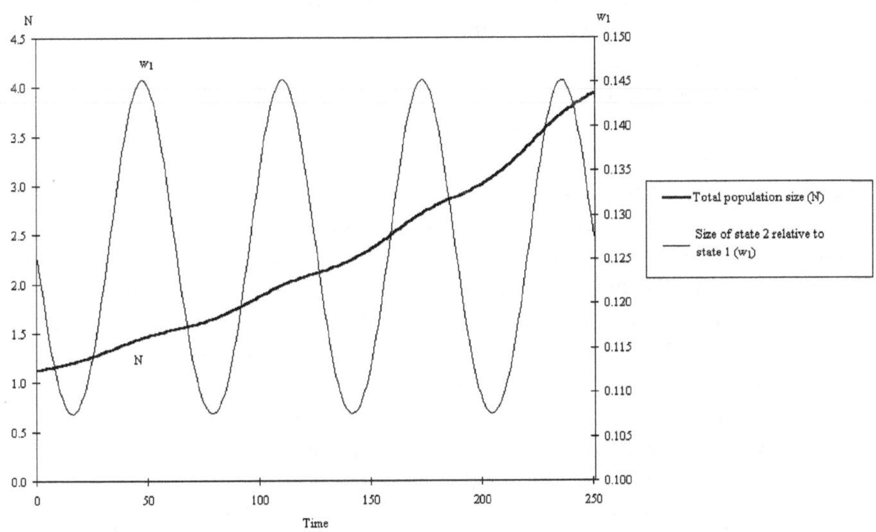


*Source:* Schoen (2001, Figure 2).

exponentially increasing trajectory. The state composition reflects the sinusoidal transfer rate from state $\underline{2}$ to state $\underline{1}$, and cycles with period $T = 20\pi$. The dynamic model thus exhibits modest fluctuations around an exponential growth trajectory, accompanied by a state structure that varies in a pronounced sinusoidal manner.

Dynamic multistate models can be generated from the Proportional Eigenvector (hyperstable) assumption or directly from assumed latent rates. Both approaches fully describe model behavior over time, and can facilitate analyses of the sensitivity of population size and structure to changes in demographic behavior.

## 8.6 FINDING RATES CONSISTENT WITH MULTISTATE POPULATION DISTRIBUTIONS

There are many occasions when data are available on the size and composition of a multistate population at two proximate points in time, but the growth and transfer rates describing the population's behavior between those time points are unknown. Census or survey data frequently provide population distributions by marital status, labor force status, place of residence, and numerous other demographic characteristics. What is sought is a method for using those data to estimate the prevailing rates of marriage and divorce, job entry and exit, interregional migration, or other relevant demographic transfers. No unique solution is possible, as with n living states there are generally n constraints on $n^2$ unknown rates.

Estimating rate matrices from beginning of interval and end of interval data on population size and structure is considerably more difficult in the multistate case than in the multi-age case (see section 7.5). With a given base rate matrix, the Proportional Transfer Rate model cannot fit arbitrary initial and end of interval population distributions. Other methods discussed in this chapter can, but in trial calculations with well behaved data they have yielded poor results. Specifically, unsuccessful attempts were made using metastable models, IDMs with both proportional eigenvalues and constant subordinate eigenvalues, hyperstable models with proportional eigenvalues, and the Additive Eigenvector assumption. Frequently, small changes in transfer rates produced substantial changes in growth or mortality rates. It appears that none of those dynamic models reflect how demographic rates typically change.

Two approaches do yield reasonable results: Iterative Proportional Fitting (IPF) and the Relative State Attraction (RSA) method. Drawing on Schoen and Jonsson (2003b), this section discusses those approaches, their conceptual underpinnings, and the rate estimates they produce.

### 8.6.1 Estimating Multistate Arrays Through Iterative Proportional Fitting

IPF, the most commonly used approach, has been known by many names, including the Deming-Stephan procedure, the DSF procedure (referring to Deming, Stephan, and Furness), bi- (or multi-) proportional adjustment, and the RAS method. The procedure's development and statistical properties have been discussed in Bishop, Fienberg, and Holland (1975) and Willekens (1982).

To describe the IPF procedure, consider a non-negative array, **D**, of r rows and c columns, whose (i, j)th element is $d_{ij}$. The $d_{ij}$ can be thought of as base values that represent prior knowledge of patterns of interaction or movement; if there is no such prior knowledge, each $d_{ij}$ can be set equal to 1. Let us specify $R_i$, $i = 1, r$, as the desired sum of the ith row, and $C_j$, $j = 1, c$, as the desired sum of the jth column. Kruithof (1937), Deming and Stephan (1940), and Furness (1965) have shown that there exists a unique set of row factors $f_i$, $i = 1, r$, and column factors $g_j$, $j = 1, c$, that produce an array $\mathbf{D}^*$ with (i,j)th element

$$d^*_{ij} = f_i \, g_j \, d_{ij} \tag{8.73}$$

that has the desired row and column totals. The $d^*_{ij}$ and the row and column factors are determined by a 3 step iterative procedure: (1) each **D** matrix element in a given row is multiplied by a row factor to produce the desired row total, and that process is repeated for every row; (2) each (adjusted) matrix element in a given column is multiplied by a column factor to produce the desired column total, and that process is repeated for every column; and (3) the row and column adjustments are continued until both row and column sums equal the desired totals. The algebraic solution is complex even for matrices of 2 rows and 2 columns, but the IPF procedure can easily be programmed for matrices of any size and convergence to a unique solution is assured.

The IPF procedure has many desirable properties. It can be applied to matrices of any size, and readily accommodates "structural zeros" (i.e. values that must be zero because a transfer between those states is not possible). Willekens (1982) showed that IPF is equivalent to estimating an array by a log linear model, when that model ignores higher order interactions. IPF is equivalent to entropy maximization (Willekens, 1999). Entropy reflects the amount of randomness (or lack of structure) in the data [cf. equations (2.17), (5.24), and the associated discussions of entropy with regard to fertility and mortality]. The maximum entropy solution essentially finds the pattern of flows that can be achieved in the greatest number of ways (Halli and Rao, 1992). Entropy maximization is especially appropriate when the probability model underlying the data is not known, as it yields estimates equivalent to those of maximum likelihood procedures (Batty and Mackie; 1972 Bishop, Fienberg, and Holland 1975, Chs 3 and 5).

Iterative proportional fitting has been widely used to estimate spatial interaction flows and input-output models. In demography, it has frequently been employed to estimate migration arrays (Chilton and Poet 1973; Nair 1985; Philipov 1978; Willekens 1982). McFarland (1975) applied IPF to the "two sex problem" of demography, using it to reconcile inconsistencies between the number of marriages implied by male and female marriage rates. Overall, reasonable results have been reported, though some migration flows involving relatively few persons were poorly estimated.

IPF is not without some shortcomings. The IPF approach to demographic rate estimation lacks strong empirical verification, as most previous demographic analyses did not compare their estimated values to known data. Conceptually, demography emphasizes the regularities that characterize population behavior, while IPF maximizes randomness. For demographic analyses, there is thus reason to consider alternative approaches based on plausible patterns of behavior.

### 8.6.2 Estimating Multistate Rates By the Relative State Attraction (RSA) Method

A simple behavioral notion is that some demographic states "attract" people while others "repel" them. For example, if marriage is perceived as less attractive, marriage rates are likely to decline and divorce rates are likely to rise. Similarly, if region R is experiencing economic prosperity, that region is apt to attract more inmigrants and experience less outmigration. While it clearly oversimplifies reality, the attraction/repulsion notion offers a plausible, behaviorally based criterion for adjusting demographic rates. Given that only n constraints are available for the estimation of $n^2$ unknown rates, any criterion that permits estimation must impose substantial restrictions.

At the core of the RSA method, which was advanced in Schoen and Jonsson (2003b), are adjustment factors that reflect the relative attraction of each living state. Let $m_{ij}(x, u)$ represent a base, or standard, rate of transfer from state i to state j between the ages of x and $x+u$. Let the age-state-specific attraction adjustment factor $k_i(x, u)$ reflect, for state i between the ages of x and $x+u$, the change in state attraction from base rate conditions. The adjusted transfer rate, $m^*_{ij}(x, u)$, is then given by

$$m^*_{ij}(x, u) = m_{ij}(x, u)[k_j(x, u)/k_i(x, u)] \qquad (8.74)$$

As the attraction of destination state j increases, the estimated rate of movement to j increases, and as the attraction of origin state i rises, the estimated rate of exit from i declines. The origin and destination state attraction factors thus interact to produce the estimated rates. The adjustment factors are symmetrical in that equation (8.74) implies

$$m^{*}_{ij}(x, u)\ m^{*}_{ji}(x, u) = m_{ij}(x, u)\ m_{ji}(x, u) \qquad (8.75)$$

For all states i and j, the product of the two adjusted transfer rates between those states equals the product of the two standard transfer rates. When rates are adjusted upwards in one direction, they are adjusted downwards in the other. Substantively, that is plausible as, for example, it is reasonable to expect an increase in divorce rates when remarriage rates fall, and for migration into a region to decline as migration out of that region rises. Such a pattern has often been observed, though exceptions are common.

To implement the RSA method, flow equations incorporating rates, derived from equations (1.46) and (1.50), are employed to link the beginning and ending populations. Let $\boldsymbol{\ell}(\mathbf{t})$ be an n element column vector describing the observed initial population, with jth element $\ell_j(t)$ representing the number of persons in state j at time t. Vector $\boldsymbol{\ell}(\mathbf{t}+\mathbf{u})$ is the analogous vector for the observed end of interval population. Person-year vector $\mathbf{L(t,u)}$, with jth element $L_j(t, u)$, represents the number of person-years lived in state j between times t and $t+u$. It can be linked to the population vectors by the linear relationship

$$\mathbf{L(t, u)} = (u/2)[\boldsymbol{\ell}(\mathbf{t}) + \boldsymbol{\ell}(\mathbf{t}+\mathbf{u})] \qquad (8.76)$$

or by an alternative person-year relationship.

Base rate matrix $\mathbf{M}$ can describe demographic behavior in any multistate population, and arrays the $m_{ij}$ in a manner very similar to base rate matrix $\boldsymbol{\mu}$ in equation (8.1). The (i,j)th element of $\mathbf{M}$, $i \neq j$, is again $m_{ji}$, but the jth diagonal element of $\mathbf{M}$ is slightly different from that in equation (8.1). It is given by

$$M_j = \phi_j - m_{j\delta} - \Sigma_i\ m_{ji} \qquad (8.77)$$

where the sum over i spans all other living states, $\phi_j$ represents the fertility rate in state j, and $m_{j\delta}$ represents the death rate in state j. The base rate matrix is best taken from a known population whose behavior is close to that of the population whose rates are being estimated.

Let matrix $\mathbf{M^*(t, u)}$ denote the array of adjusted rates. The (i,j)th element of $\mathbf{M^*(t, u)}$, $i \neq j$, is $m^*_{ji}(t, u) = m_{ji}(t, u)[k_i(t, u)/k_j(t, u)]$. The jth diagonal element is

$$M^*_j(t, u) = \phi_j(t, u)\ k_j(t, u) - m_{j\delta}(t, u)/k_j(t, u) - \Sigma_i m^*_{ji}(t, u) \qquad (8.78)$$

The fertility rate in state j is adjusted proportionally to the state j adjustment factor, while the mortality rate in state j is adjusted inversely to that factor. Essentially, the "dead" and "preborn" states are assumed to have adjustment factors of 1. [Alternatively, additional data on mortality and fertility can be brought to bear. Schoen and Jonsson (2003, note 3) discuss incorporating such additional data.]

Adjusted rate matrix $\mathbf{M}^*(\mathbf{t}, \mathbf{u})$ must satisfy the multistate matrix flow equation

$$\boldsymbol{\ell}(\mathbf{t}+\mathbf{u}) = \boldsymbol{\ell}(\mathbf{t}) + \mathbf{M}^*(\mathbf{t}, \mathbf{u})\ \mathbf{L}(\mathbf{t}, \mathbf{u}) \qquad (8.79)$$

The adjustment factors are found by solving the set of scalar equations embodied in (8.79). Those equations are nonlinear, and even the general two living state model gives rise to a complicated cubic solution. Numerically, the solution can readily be found using a mathematical program such as Maple, Mathematica, or S+. Numerous calculations involving models with 2, 3, and 4 living states have been made, and all have yielded only one solution where every $k_j(t, u) > 0$. When the equations include mortality or fertility, a numerical value for each $k_j(t, u)$ can be obtained. When there is no fertility or mortality in the model, one equation in (8.79) is linearly dependent on the other $(n - 1)$, and the solution is best obtained in terms of $k_j(t, u)/k_1(t, u)$ ratios. Exercise 12 shows the algebraic solution for the two living state model without fertility and mortality, and demonstrates that there is only one solution where $k_2(t, u)/k_1(t, u) > 0$.

The RSA method has a number of strengths. It is based on a simple and reasonable behavioral premise that can readily be explained to nontechnical audiences. It is applicable to models of any size, easily accommodates structural zeros, and is straightforward to apply in practice. The RSA approach also allows some analytical options that are not available with IPF. Because RSA is based on an interpretable behavioral assumption, it can be used to examine scenarios expressed in terms of changes in state attraction. For example, one can begin with an observed population and set of rates, and assume that the attraction of given states changes in specified ways over time. RSA calculations can show the population trajectories that follow from those changes in state attraction. Such scenarios cannot be developed using IPF, because that procedure is not able to incorporate patterns of change based on behavioral assumptions.

RSA has its limitations as well. As a new method, it is less familiar to potential users and lacks a track record. Its statistical properties are not known. Moreover, it has not been proven that there is always one and only one solution that yields positive adjustment factors. The existence of such a unique solution may be dependent on the choice of standard rate matrix $\mathbf{M}$.

### 8.6.3 Evaluating the IPF and RSA Approaches

Schoen and Jonsson (2003b) evaluated the IPF and RSA approaches by comparing their estimates to known values. Two of their comparisons systematically changed the base rates, and a third compared an estimated rate multistate model with one calculated from observed data. That study appears to provide the only systematic analysis of either method where the estimates are compared to known demographic rates.

Indirect standardization is a demographic technique that exploits the typical regularity in demographic behavior over age (cf. Preston, Heuveline and Guillot 2001). Starting with that idea, the first two comparisons multiplied age schedules of observed rates, at all ages, by different factors, and examined how the RSA and IPF estimates responded to those new base rates. That imposed pattern of change is not related to the assumptions underlying either estimation procedure.

The first comparison was based on the two living state (Married/Unmarried) model for Sweden Females Born 1930–34, which was presented in Schoen (1988a, p92–3). Published life table survivorship values at ages x and $x+5$ provided the initial and final population figures for both states. The evaluation focused on ages 15–39, and fertility rate $\phi_j$ was set to zero. For IPF, estimation was based on equation (8.73). For RSA, equation (8.79) was used to determine the $k_j(x, 5)$, and the $m^*_{ij}(x, 5)$ followed from them using equations (8.74) and (8.78).

Figure 8.7 shows 4 sets of marriage (U to M) rates: actual, estimated by IPF, estimated by RSA, and the "standard" rates obtained by multiplying the observed rates by factors of 0.7 and/or 1.4. In Panel A, observed rates from M to U and from U to M were both multiplied by 0.7 and used as the standard rates. The estimates from both procedures are reasonable at all ages and follow the actual age pattern quite well, though they are somewhat low at ages 25 through 59. The RSA and IPF estimates are extremely similar. In Panel B, the observed rates from M to U were multiplied by a factor of 1.4 while the observed rates from U to M were multiplied by a factor of 0.7. Since the product of those factors is close to 1, those adjustments are consistent with the symmetrical relationship underlying the RSA method and expressed in equation (8.75). Panel B shows that both estimates are extremely close to the actual values, with the only palpable difference appearing in the large rate for the 25–29 year age group. Panel C shows the results when both observed rates were multiplied by factors of 1.4. Both estimates remain reasonable, though there is a sizeable discrepancy at age 25–29 and the estimates are somewhat high at ages 20 through 54. In every case, at almost every age, the RSA and IPF estimates are nearly indistinguishable.

Estimates were also made using factors of 0.8, 0.9, 1.1, and 1.25, and the Married to Unmarried rates were examined as well. Those results reinforced

*Figure 8.7. Actual, Estimated, and Standard Marriage Rates From a Two-Living State (Married/Unmarried) Model Based on Rates for Swedish Women, Cohort Born 1930–1934*

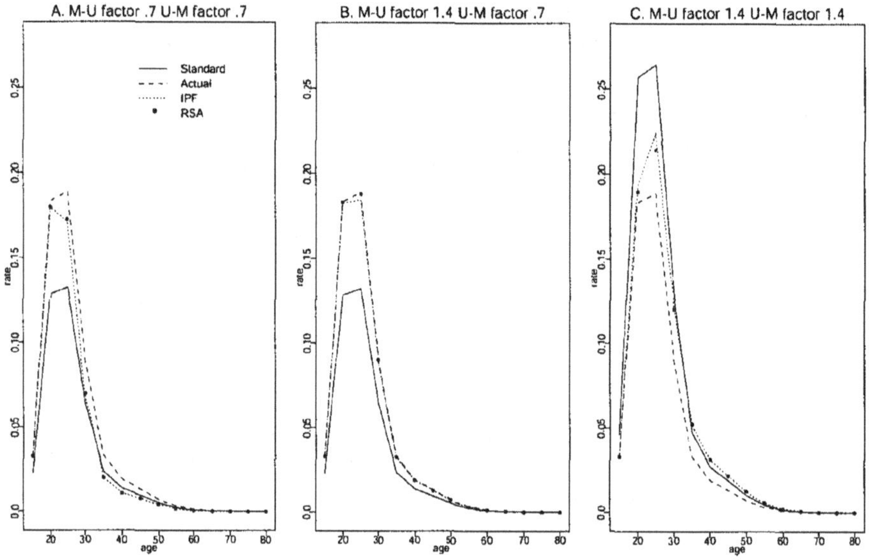


*Source:* Schoen and Jonsson (2003b, Figure 1).

the pattern shown in Figure 8.7. Errors of estimation are quite small when the product of the adjustment factors is close to 1, but increase when that product diverges from 1. The age pattern of the estimates consistently parallel the actual rates, and the RSA and IPF estimates are invariably close.

A similar evaluation considered the 4 living state (Never Married, Presently Married, Widowed, and Divorced) marital status life table for United States Females 1995 examined in Schoen and Standish (2001). The adjustment factors again ranged from 0.7 to 1.4, and were applied to the first marriage and divorce rates. The effects of different combinations of those factors on estimated rates of first marriage, divorce, and remarriage from divorce were then examined. Since there are no transfers to the Never Married state, estimates of first marriage rates are consistently good. Estimates of the other two rates followed a pattern similar to that in Figure 8.7. The closeness of the estimates varies with the deviation of the product of the adjustment factors from 1, though the age pattern of the estimated rates is consistently reasonable and the RSA and IPF estimates are always very similar.

A third evaluation was performed using the same 4 living state model. It compared actual and estimated rates for US Females 1995 taking observed rates for US Females 1988 as the standard. Although the 1988 rates are a reasonable choice, U.S. age-specific marriage and divorce patterns changed significantly over that interval. In 1995, first marriage, divorce, and remarriage rates were substantially lower at younger ages but somewhat higher at older ages than they were in 1988. As a result, the product of the implicit rate adjustment factors is not close to 1. The resultant pattern of the estimated rates consequently resembles the results obtained when the standard 1995 rates were manipulated by arbitrary factors. First marriage rate estimates are close, while divorce and remarriage rate estimates are somewhat off. There were large errors in estimated rates of remarriage from divorce at ages under 25, as both transfer rates fell and there was

*Table 8.1. Selected Summary Measures from Actual 1988, Actual 1995, and Estimated 1995 Female Marital Status Life Tables for the United States*

| Measure | Actual | | RSA Estimated | IPF Estimated |
|---|---|---|---|---|
| | 1988 | 1995 | 1995 | 1995 |
| 1. Proportion ever marrying of those surviving to age 15 | .879 | .887 | .886 | .884 |
| 2. Mean age at first marriage | 25.1 | 26.6 | 26.6 | 26.6 |
| 3. Proportion of marriages ending in divorce | .432 | .425 | .415 | .419 |
| 4. Mean age at divorce | 34.4 | 37.3 | 35.5 | 35.6 |
| 5. Average duration of a marriage | 24.8 | 25.7 | 25.8 | 25.8 |
| 6. Remarriages of widowed persons per widowhood | .063 | .048 | .057 | .056 |
| 7. Remarriages of divorced persons per divorce | .723 | .687 | .695 | .695 |
| 8. Mean age at remarriage from divorce | 36.0 | 39.7 | 37.8 | 37.8 |

*Source:* (Schoen and Jonsson, 2003b, Table 4).

a particularly large 1988–1995 decline at those ages in rates of remarriage from divorce. Estimates from the RSA and IPF methods are again quite close, and IPF estimates again worsen when the assumptions underlying the RSA method are violated. Overall, Table 8.1 shows that the errors in the rate estimates do not substantially effect most major summary measures of marriage and divorce, though the estimated mean ages at divorce and at remarriage from divorce are off by nearly 2 years. Both estimation methods thus seem reasonably robust to fairly considerable departures from the behavior described by the standard rates.

In sum, there appear to be two viable approaches to estimating an array of multistate transfer rates given the initial and end of interval populations in each state. The IPF procedure has a stronger statistical foundation, while the RSA approach is based on a simple and plausible behavioral assumption. Given their conceptual and mathematical differences, it is striking that the two approaches consistently yield very similar estimates. Both procedures are standard dependent. Estimates worsen when the shift in the observed rate from state i to state j, relative to its standard rate, is not offset by a shift in the opposite direction on the part of the observed rate from state j to state i, relative to its standard rate. Nonetheless, it is reassuring that estimates based on maximum likelihood/entropy maximization and estimates based on changes in relative state attraction bear a great resemblance to one another.

## 8.7 SUMMARY

Dynamic models without age but with more than one living state provide new analytical opportunities and additional complexities. In cohort analyses where time and age are the same, the proportional transfer rate (PTR) assumption allows risks of interstate movement to be aggregated over age, so that state composition at every age can be determined analytically. The PTR assumption is tenable for a number of demographic behaviors, and applications of PTR models to analyses of parity progression and marriage/divorce/remarriage are discussed.

Both the metastable and IDM approaches can be extended to the multistate case. A number of interesting relationships have emerged, but little applied work using such models has been done to date. IDMs with uniform growth across all living states are examined in some detail, with particular attention given to the analysis of two living state models. Multistate hyperstable models are then considered, and their relationships to "latent rates" and to birth-death models are explored, especially in two living state models.

None of the dynamic models considered provide reasonable estimates of multistate transfer rates over an interval from data on the beginning and end of interval populations in each state. Two methods that do, Iterative Proportional Fitting (IPF) and Relative State Attraction (RSA), are discussed. Despite great differences in their underlying assumptions, those methods consistently yield very similar estimates.

## 8.8 EXERCISES

[See Appendix B for selected answers]

1. Let rate matrix $\boldsymbol{\mu}$ be defined as in equation (8.66). Show that the associated population projection matrix can be written

$$\mathbf{A} = \begin{bmatrix} \frac{-R_1(r_2-\rho_1+m_{12})+R_2(r_1-\rho_1+m_{12})}{(r_1-r_2)} & \frac{m_{21}(R_1-R_2)}{(r_1-r_2)} \\ \frac{m_{12}(R_1-R_2)}{(r_1-r_2)} & \frac{R_1(r_1-\rho_1+m_{12})-R_2(r_2-\rho_1+m_{12})}{(r_1-r_2)} \end{bmatrix}$$

where the eigenvalues of $\mu$ are $r_{1,2} = [\rho_1 - m_{12} + \rho_2 - m_{21} \pm \{[\rho_1 - m_{12} - \rho_2 + m_{21}]^2 + 4m_{12}m_{21}\}^{1/2}]/2$ and $R_j = \exp[n\, r_j]$, with n the length of the age/time interval.

2. Consider a 4 living state hierarchical PTR with a $\boldsymbol{\mu}$ matrix in the form of equation (8.17). Verify that the right eigenvector matrix of $\boldsymbol{\mu}$ is given by

$$\mathbf{U} = \begin{bmatrix} -ABC/\mu_{12}\mu_{23}\mu_{34} & 0 & 0 & 0 \\ BC/\mu_{23}\mu_{34} & DE/\mu_{23}\mu_{34} & 0 & 0 \\ -C/\mu_{34} & -E/\mu_{34} & -F/\mu_{34} & 0 \\ 1 & 1 & 1 & 1 \end{bmatrix}$$

where $A = (\mu_{12} - \mu_{23})$, $B = (\mu_{12} - \mu_{34})$, $C = (\mu_{12} - \mu_{45})$, $D = (\mu_{23} - \mu_{34})$, $E = (\mu_{23} - \mu_{45})$, and $F = (\mu_{34} - \mu_{45})$.

3. Consider the hierarchical PTR parity progression model of equation (8.17) in the special case where all transfer rates are equal. With $y_j(x)$ the proportion in state j (parity $j-1$) at age x, Z(x) the sum of the transfer rates to parity 1 up to age x, and the age 0 population consisting of one person in state 1, show that
   a. $y_j(x) = Z^{j-1}(x)\ \exp[-Z(x)]/(j-1)!$
   b. $\Sigma_j\ y_j(x) = 1$ at all ages
   c. when $Z(x) = (j-1)$, $y_j(x)$ is at its maximum value [e.g. the maximum fraction of the cohort in parity 3 ($j = 4$) is achieved when $Z(x) = 3$]
   d. the maximum value of $y_j(x)$ in part c is given by $(j-1)^{j-1}\exp[-(j-1)]/(j-1)!$

4. Consider the PTR marriage/divorce/remarriage model whose state distribution is given in equation (8.23). Show that if $[\mu_{mv} + \mu_{vm}] > \mu_{sm} > \mu_{vm}$ there is a value of Z at which the proportion married attains a maximum value that is greater than its ultimate stable value.

5. Let the sequence of two living state PPMs $\mathbf{A_t}$ have the form

$$\mathbf{A_t} = \lambda_t \begin{bmatrix} 1 - u_t\ v_t & v_t \\ u_t(1 - u_t\ v_t) & u_t\ v_t \end{bmatrix}$$

Show that the product matrix from time 0 to time t, $\mathbf{M_{0,t}} = \mathbf{A_t}\ \mathbf{A_{t-1}} \ldots \mathbf{A_1}$, can be written

$$\mathbf{M_{0,t}} = Q^*\{\Pi\lambda_j\}\mathbf{u_t}$$

where the product is taken over j from 1 to t, $Q^*$ is a scalar, and the transpose of $\mathbf{u_t}$, $\mathbf{u_t}' = [1\ u_t]$. Determine the form of $\mathbf{v_t}'$.

6. Consider the multistate metastable model of equation (8.24) when $\mathbf{H}$ is a diagonal matrix whose (1,1) element is 1 and whose jth diagonal element is $h_j$. Let the time t dominant right eigenvector of $\mathbf{A_t}$ be $\mathbf{u_t}$ and the time t dominant left eigenvector of $\mathbf{A_t}$ be $\mathbf{v_t}'$. Verify that
a. the jth element of $\mathbf{u_t}$, $u_{jt}$, is given by $u_{jt} = u_{j0}\ h_j^{\ t}$
b. the jth element of $\mathbf{v_t}'$, $v_{jt}$, is given by $v_{jt} = v_{j0}/h_j^{\ t}$
c. with 1 person in state 1 at time 0, the time t stable equivalent number of persons in state 1 of the metastable population is

$$Q_t = \lambda_s^{\ t}\Sigma_j\ u_{js}\ v_{j0}$$

where the sum ranges over all states, $\lambda_s$ is the dominant root of metastable synthesis matrix $\mathbf{S}$, and $u_{js}$ is the jth element of the dominant eigenvector of $\mathbf{S}$.
d. the metastable state-specific momentum for state x at time t is given by the time invariant expression

$$\Omega_{xt} = u_{x0}\Sigma_j\ u_{js}\ v_{j0}/u_{xs}$$

e. with $Q_{xt}$ the contribution to $Q_t$ from state x at time t and $q_{xt} = Q_{xt}/Q_t$, we have the time invariant relationship

$$q_{xt} = [u_{xs}\ v_{x0}]/\Sigma_j\ u_{js}\ v_{j0}$$

f. the time invariant value of the Kullback distance between the time t multistate metastable population and its associated stable population is given by

$$K(t) = -\Sigma_x\{[u_{xs}\ v_{x0}]/\Sigma_j\ u_{js}\ v_{j0}]\}\{\ln\ [u_{x0}\Sigma_j\ u_{js}\ v_{j0}/u_{xs}]\}$$

7. In the no growth, constant metabolism model of equations (8.30) and (8.34), show that
a. the time t Kullback distance between the model population and its associated stable population is given by

$$K(t) = \ell_1(t)\ \ln[\ell_1(t)\ M/m_{21}(t)] + \ell_2(t)\ \ln[\ell_2(t)\ M/\{M - m_{21}(t)\}]$$

b. $K(t) = 0$ when $\ell_1(t) = m_{21}(t)/M$, i.e. when the $\ell_1(t)$ and $m_{21}(t)/M$ curves intersect
8. Consider the no growth, constant metabolism model of equations (8.30) and (8.34). Show that
a. if $t_{ext}$ is the time at which an extremum of $\ell_1(t)$ occurs,

$$t_{ext} = \tan^{-1}(-M/\omega)/\omega$$

b. the first extremum of $\ell_1(t)$ is a maximum and occurs between times 3T/4 and T years.
c. at time $t_z$, when $\ell_1(t_z) = m_{21}(t_z)/M$, $\tan\ \omega t_z = -M/\omega$, hence $\ell_1(t_z)$ is at an extremum
d. let the relative amplification, RA, the ratio of $\ell_1(t_{ext})$ to $m_{21}(T/4)/M$ be given by

$$RA = [1 + b\ \sin(\omega t_{ext})]/[1 + b]$$

Show that implies that the maximum of $\ell_1(t)$ is less than the maximum of $m_{21}(t)/M$, and that the minimum of $\ell_1(t)$ is greater than the minimum of $m_{21}(t)/M$.
9. Consider the constant metabolism, no growth model of equation (8.36). Show that
a. if $m_{21}(t) = m_{21}(0)\ e^{at}$, as in equation (8.31), then equation (8.39) yields

$$1/\mu_{Mt} = [m_{21}(0)\ e^{at}(1 - e^{-M})]/[M(1 - e^{-M-a})]$$

b. show that under the exponential rate function of equation (8.31), equation (8.32) can be rewritten $\ell_2(t)/\ell_1(t) = \mu_{Mt} - 1 = [M + a - m_{21}(0)\ e^{at}]/[m_{21}(0)\ e^{at}]$
c. reconcile the result for $\ell_2(t)/\ell_1(t) = (\mu_{Mt} - 1)$ in b with the result for $\mu_{Mt}$ in a.
10. Consider the $\boldsymbol{\mu}$ matrix of equation (8.21). Show that if $\mu_{sm}$ and $(\mu_{mv} + \mu_{vm})$ are constant over time, $\boldsymbol{\mu}_t$ is an IDM no growth rate matrix with one subordinate right eigenvector that varies over time.

11. In the Additive Eigenvector approach, matrix **W** of equation (8.52) satisfies the relationship

$$\mathbf{W(t)} = \mathbf{J(t)} + \mathbf{K}$$

where **J(t)** is an $n \times n$ matrix with ones in the first row and $J_h(t) = w_{h1}(t)$ as the value of all elements in the hth row, and **K** is a constant $n \times n$ matrix whose first row and first column elements are all zero and whose (i,j)th element is $k_{ij}$. Show that in the two living state case, the Additive Eigenvector assumption leads to the solution in equation (8.65) with $d^* = d$.

12. A non-hierarchical two living state RSA model with no fertility and no mortality can be described by the equations

$$\ell_1(t+u) = \ell_1(t) - Z\, L_1(t,u)\; m_{12}(t,u) + L_2(t,u)\; m_{21}(t,u)/Z$$

$$\ell_2(t+u) = \ell_2(t) - L_2(t,u)\; m_{21}(t,u)/Z + Z\, L_1(t,u)\; m_{12}(t,u)$$

where $Z = k_2(t, u)/k_1(t, u)$.

a. Show that those equations lead to the quadratic solution

$$Z = ([\ell_1(t) - \ell_1(t+u)] \pm [\{\ell_1(t) - \ell_1(t+u)\}^2 + 4\, L_1(t,u)m_{12}(t,u)L_2(t,u)m_{21}(t,u)]^{1/2})/[2L_1(t,u)m_{12}(t,u)]$$

b. Show that the positive root in the above solution always yields a positive Z and the negative root a negative Z, proving that there is one and only one positive solution.

13. Let $\mathbf{B_t}$ designate a sequence of multistate PPMs, and let

$$\mathbf{A_t} = \mathbf{B_t B^{-1}{}_{t-1}}$$

If $\mathbf{M_{0,t}} = \mathbf{A_t\ A_{t-1} \ldots A_1}$, show that

$$\mathbf{M_{0,t}} = \mathbf{B_t B^{-1}{}_0}$$

indicating that the product matrix of such an $\mathbf{A_t}$ sequence is readily found.

14. Let PPM $\mathbf{A_t}$ be of the form

$$\mathbf{A_t} = \mathbf{K^{f(t)} C_t K^{-f(t)}}$$

where $\mathbf{K}$ is a diagonal matrix with nonnegative elements, $\mathbf{C}_t$ a sequence of multistate PPMs, and f(t) a specified function of time. Show that if

$$f(t) = t + ht^2$$

and

$$\mathbf{C}_t = \mathbf{C}_1\mathbf{K}^{2h(t-1)}$$

then $\mathbf{M}_{0,t} = \mathbf{A}_t\ \mathbf{A}_{t-1} \ldots \mathbf{A}_1$ is given by

$$\mathbf{M}_{0,t} = \mathbf{K}^{t+h(t\wedge 2)}(\mathbf{C}_1\ \mathbf{K}^{-1-h})^t$$

where $(t\wedge 2) = t^2$. Confirm that the time varying $\mathbf{C}_t$ allow a variant of the metastable approach to describe growth that does not follow a simple exponential.

15. Consider a PPM $\mathbf{A}$ with dominant root $\lambda$, and let $\mathbf{W}$ be a diagonal matrix whose inverse is $\mathbf{Z}$ and whose jth diagonal element is the jth element of the dominant right eigenvector of the PPM. Show that the product $\mathbf{ZAW}/\lambda$, an extended Sykes transformation, yields a matrix whose rows sum to 1.

# CHAPTER 9

# DYNAMIC MODELS WITH MULTIPLE AGES AND STATES

## 9.1 INTRODUCTION

Models of demographic renewal and of movement between different living states have proven to be of great analytical value. Allowing vital rates to change over time enhances that value. Combining renewal, interstate movement, and change over time leads to dynamic models incorporating both age and state. By generalizing the multi-age, multistate stable populations of Section 1.7, such models integrate the multi-age approach of Chapter 7 with the multistate emphasis of Chapter 8. They allow the three core processes of demography—mortality, fertility, and migration—to be analyzed simultaneously in the context of populations with changing rates. Dynamic multi-age and multistate models are in their infancy, however, and working with them can involve both extensive data inputs and analytical complexity.

The easiest way to approach models with multiple ages and states is through the use of discrete matrix models. Such matrices can be written in several ways, but we follow Feeney (1970) and use the "block Leslie" approach introduced in equation (3.24). Whenever data are available, discrete multi-age and state models can be projected over time using conventional projection (or matrix multiplication) methods. That approach can be quite useful, as in the United Nations multiregional population projections (e.g. United Nations 2005).

In this chapter, we focus on extensions of the dynamic approaches discussed earlier. The hyperstable model is considered first, with an application involving the health statuses Robust and Frail. The metastable model, whose specification requires an additional step, is then considered. As of now, no way has been found to extend intrinsically dynamic models to the multi-age, multistate case.

## 9.2 THE MULTI-AGE AND STATE HYPERSTABLE (MASH) MODEL

### 9.2.1 Specifying the MASH Model

The hyperstable model of Subsection 7.5.1 and the PTR model of Section 8.2 can be combined to provide a flexible platform for analyzing populations with multiple ages and states. Essentially, the MASH model uses the hyperstable approach to vary age-state-specific fertility and the proportional transfer rate assumption to survive each cohort. Given the model's complexity, it is most conveniently expressed in terms of matrices whose elements are matrices. To facilitate the presentation, we assume a model with three 15-year age groups and two living states, but the approach taken can readily be applied to any number of ages and states.

The MASH matrix that projects the time t-1 population, at all ages, to time t, is denoted $\mathbf{A}_t$. That PPM is expressed as the product

$$\mathbf{A}_t = \wp_t\, \mathbf{X}_t\, \mathcal{F}\, \mathbf{X}^{-1}{}_{t-1}\, \wp^{-1}{}_{t-1} \tag{9.1}$$

Matrix $\mathcal{F}$ is the base block Leslie matrix. The diagonal $\mathbf{X}$ matrices determine the size and allocation of birth cohorts, and the block diagonal $\wp$ matrices survive (or advance) birth cohorts over age and time. To fully define each of those matrices, we start at the center and work outward.

Matrix $\mathcal{F}$ is a time invariant, stationary, block Leslie matrix. It is analogous to the $\mathbf{F}$ matrix defined in equation (7.25), but with matrix elements. Specifically,

$$\mathcal{F} = \begin{bmatrix} \mathbf{F}_1 & \mathbf{F}_2 & \mathbf{F}_3 \\ \mathbf{I} & \mathbf{0} & \mathbf{0} \\ \mathbf{0} & \mathbf{I} & \mathbf{0} \end{bmatrix} \tag{9.2}$$

where $\mathbf{I}$ is the $2 \times 2$ identity matrix, $\mathbf{0}$ a $2 \times 2$ matrix all of whose elements are zero, and $\mathbf{F}_i$ is a $2 \times 2$ matrix of age-state-specific fertility rates, that is

$$\mathbf{F}_i = \begin{bmatrix} f_{i11} & f_{i21} \\ f_{i12} & f_{i22} \end{bmatrix} \tag{9.3}$$

with $f_{ijk}$ denoting the number of persons in age group 1, state k, at the end of the interval per person in age group i, state j, at the beginning of the interval. Matrix $\mathcal{F}$ is defined to be row stochastic, i.e. the sum of the kth row elements over all of the $\mathbf{F}_i$ is 1.

Diagonal matrix $\mathbf{X}_t$ reflects the number and allocation of births, i.e. persons in the first age group, at time t and earlier times. Let the scalar $x_t$ be the total number of persons in the first age group, in all states, at time t, and let $\mathbf{G}_t$ be the $2 \times 2$ diagonal matrix

$$\mathbf{G}_t = \begin{bmatrix} 1 - g_t & 0 \\ 0 & g_t \end{bmatrix} \tag{9.4}$$

where $g_t$ represents the fraction of time t births in state 2. We can then write

$$\mathbf{X}_t = \begin{bmatrix} x_t\mathbf{G}_t & \mathbf{0} & \mathbf{0} \\ \mathbf{0} & x_{t-1}\mathbf{G}_{t-1} & \mathbf{0} \\ \mathbf{0} & \mathbf{0} & x_{t-2}\mathbf{G}_{t-2} \end{bmatrix} \tag{9.5}$$

Block diagonal matrix $\wp_t$ projects (i.e. advances) every cohort living at time $t-1$ forward to time t with respect to both survivorship and interstate transfer. It can be written

$$\wp_t = \begin{bmatrix} \mathbf{I} & \mathbf{0} & \mathbf{0} \\ \mathbf{0} & \mathbf{P}_{2t} & \mathbf{0} \\ \mathbf{0} & \mathbf{0} & \mathbf{P}_{3t}\,\mathbf{P}_{2,t-1} \end{bmatrix} \tag{9.6}$$

where $\mathbf{P}_{jt}$ is the age-time-specific matrix that advances those in age group $j-1$ at time $t-1$ to age group j at time t. So that behavior can be readily cumulated over age intervals, each cohort advances subject to a specified, constant PTR base matrix. The base matrix for the cohort in age group 1 at time t is denoted by П, the Cyrillic letter P, with

$$\mathbf{\Pi}_t = \begin{bmatrix} p_{11t} & p_{21t} \\ p_{12t} & p_{22t} \end{bmatrix} \tag{9.7}$$

where $p_{jkt}$ is the number of persons in the kth state at the end of the interval per person in the jth state at the beginning of the interval. Using equations (9.6) and (9.7), we can write

$$\mathbf{P}_{jt} = [\mathbf{\Pi}_{t-j+1}]^{z(j,t)} \tag{9.8}$$

where z(j,t) reflects the scalar factor adjusting all of the underlying base transfer rates from age group $j-1$ at time $t-1$ to age group j at time t. Equation (9.8) is analogous to PTR equation (8.4). Using equation (9.8), the (3,3) element in the matrix in equation (9.6) can be written

$$\mathbf{P}_{3t}\,\mathbf{P}_{2,t-1} = [\mathbf{\Pi}_{t-2}]^{z(3,t)+z(2,t-1)} \tag{9.9}$$

If there are more than 3 age intervals, longer products of **P** matrices will arise, but every such product can be represented as a single **Π** matrix raised to a power. Employing equations (9.1) through (9.8), MASH PPM $\mathbf{A}_t$ can be written

$$\mathbf{A}_t = \begin{bmatrix} x_t\mathbf{G}_t\mathbf{F}_1(x_{t-1}\mathbf{G}_{t-1})^{-1} & x_t\mathbf{G}_t\mathbf{F}_2(x_{t-2}\mathbf{G}_{t-2})^{-1}\mathbf{P}^{-1}{}_{2,t-1} & x_t\mathbf{G}_t\mathbf{F}_3(x_{t-3}\mathbf{G}_{t-3})^{-1}\mathbf{P}^{-1}{}_{2,t-2}\mathbf{P}^{-1}{}_{3,t-1} \\ \mathbf{P}_{2t} & \mathbf{0} & \mathbf{0} \\ \mathbf{0} & \mathbf{P}_{3t} & \mathbf{0} \end{bmatrix} \tag{9.10}$$

The long term population vector at time t has the form

$$\mathbf{x}_t = \begin{bmatrix} \mathbf{g}_t\ x_t \\ \mathbf{P}_{2t}\ \mathbf{g}_{t-1}\ x_{t-1} \\ \mathbf{P}_{3t}\ \mathbf{P}_{2,t-1}\ \mathbf{g}_{t-2}\ x_{t-2} \end{bmatrix} \tag{9.11}$$

where vector $\mathbf{g}_t$ is defined by

$$\mathbf{g}_t = \begin{bmatrix} 1 - g_t \\ g_t \end{bmatrix} \tag{9.12}$$

It follows from equations (9.10) - (9.12) that the projection relationship holds, i.e.

$$\mathbf{x}_t = \mathbf{A}_t\ \mathbf{x}_{t-1} \tag{9.13}$$

The first element of $\mathbf{x}_t$ is reproduced because $\mathbf{F}_1 + \mathbf{F}_2 + \mathbf{F}_3$ is row stochastic.

The form of equation (9.1) simplifies the product of the PPMs over time. With the product matrix $\mathbf{M}_{0,t} = \mathbf{A}_t\ \mathbf{A}_{t-1} \cdots \mathbf{A}_1$, we find

$$\mathbf{M}_{0,t} = \wp_t\ \mathbf{X}_t\ \mathcal{F}^t\ \mathbf{X}^{-1}{}_0\ \wp^{-1}{}_0 \tag{9.14}$$

In the long term, $\mathcal{F}^t$ becomes a rank one matrix, and the effects of any arbitrary initial population composition are reduced to a scalar factor. The dominant eigenvalue of $\mathcal{F}$ is 1. The dominant right eigenvector of $\mathcal{F}$, $\mathbf{u}_{\mathcal{F}}$, is simply a column vector of ones. (The dominant left eigenvector of $\mathcal{F}$, $\mathbf{v'}_{\mathcal{F}}$, is more complex, and depends on the specified age-state pattern of fertility.)

The MASH model is quite general in scope, and provides consistent, closed form expressions for the time-specific PPMs and population vectors. It can accommodate any specified birth sequence, and allows the births of each year to be allocated across states in any chosen manner. Cohort advancement must satisfy the proportional transfer rate assumption, but each cohort can have its own base transfer matrix, and a chosen adjustment factor can be applied to each age at every time. The number of person-years lived in each state during each time interval can be found from the population vectors and a person-year assumption (cf. equations (1.9) and (8.9)). The number of movements between specified states during each interval quickly follows (cf. equation (1.10)). Nonetheless, simplifying the product of the PPMs over time comes at a price. The fertility rates are determined by the choices made regarding birth cohort size, state allocation, and survivorship. The pattern of cohort advancement does not influence either birth cohort size or state composition. The projection expression for the long term model, $\mathbf{x}_t = \mathbf{A}_t\mathbf{x}_{t-1}$ where $\mathbf{A}_t$ is given by equation (9.1), essentially operates by undoing previous cohort advancement, eliminating the initial size and composition of every cohort, creating unit birth cohorts via $\mathcal{F}$, scaling and allocating those cohorts, and then surviving them to time t.

### 9.2.2 An Illustrative Application

The MASH model has a number of potential applications. It can model a wide range of scenarios, for example a fertility-mortality transition accompanied by urbanization. It allows the specification and identification of distinct age, period, and cohort effects (see Exercise 6). Wedge-period models arise when the birth cohorts are of uniform size, and can be used to examine changing patterns of survivorship and interstate transfer.

As an illustration, consider a population with two living health statuses, "Robust" and "Frail". The initial composition of each birth cohort is assumed to change linearly from 90% Frail at times 0 and earlier to 90% Robust at times 80 and later. Given a fixed base PTR advancement matrix, death and interstate movement rates vary over age, but age-state-specific mortality and age-state-specific rates of interstate transfer remain constant over time. At the level of the total population, the model describes how changing population heterogeneity affects survivorship.

Age-specific death rates for the Robust state are taken from the Female U.S. Decennial Life Tables for 1989–91 (National Center for Health Statistics 1997). At each age, the death rates in the Frail state are three times those in the Robust state. Each age-specific rate of transfer from Robust to Frail is the same as the rate from Frail to Robust, and both are set at twice the death rate in the Robust state. As a result, the base $\mu$ matrix has elements $-3$ and 2 in the first row and 2 and $-5$ in the second row. For each period, an associated life expectancy at birth was calculated using death rates for each age weighted by the period population's health status composition at that age. In each age-specific base fertility matrix, the two rows are equal. For $\mathbf{F}_1$, $\mathbf{F}_2$, and $\mathbf{F}_3$, respectively, the elements in each row are [.15, .15], [.30, .30], and [.05, .05].

Figure 9.1 shows, for years 0 through 205, the time trajectory of the percent Frail in the period population, the size of the period population (CAL(t)),

*Figure 9.1. Summary Measures of the Trajectory of a Robust/Frail MASH Model*

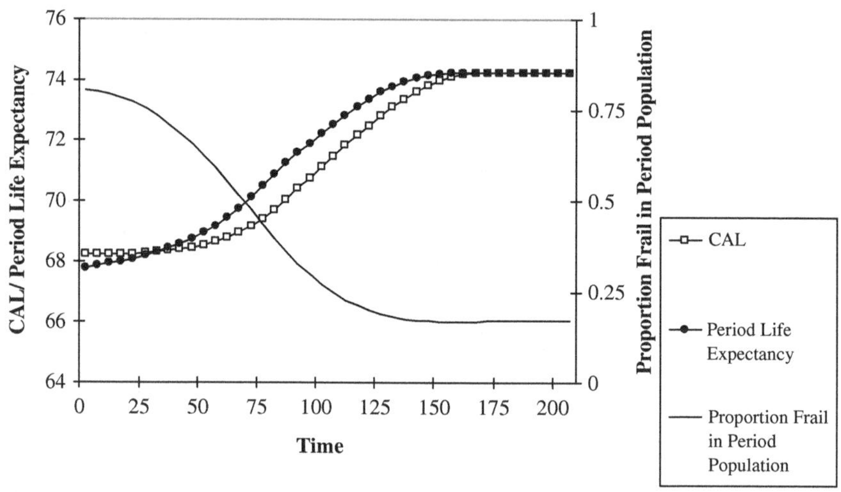


and the period life expectancy at birth (e(0,t)). Initially, the period population is a multistate stationary population based on birth cohorts 90% Frail. In year 205, it is essentially a multistate population based on birth cohorts that are 90% Robust. In the total population, the proportion Frail varies from 80.5% to 17.0%. With birth cohorts of unit size, the total number of persons in the period population goes from 68.22 to 74.24, while the associated period life expectancy varies from 67.79 to 74.23. Even at years 0 and 205, life expectancy does not equal CAL because survivorship in the associated life table does not reflect the interstate dynamics present in the multistate model. Period life expectancy increases more rapidly during the transition from birth cohorts that are 90% Frail to those 90% Robust because the decrease in frailty at birth immediately affects all ages in the associated life table.

One important caveat must be noted regarding applications of the MASH model. The form of PPM $\mathbf{A}_t$ in equation (9.10) includes inverses of advancement matrix $\mathbf{P}$. The $\mathbf{P}^{-1}$ matrices have negative elements, which makes it possible for the first row elements of MASH PPM $\mathbf{A}_t$ to be negative. In the example considered here, that would occur if a birth cohort were 95% Frail. Negative first row elements are less likely to arise when birth allocations and fertility rates are more equal across states. Negative values in $\mathbf{A}_t$ indicate that the desired birth cohort state composition could not arise in a real population that had the model population's composition and the given fertility and advancement patterns.

## 9.3 THE MULTI-AGE AND STATE METASTABLE (MASM) MODEL

The MASM model is similar to the MASH model in a number of respects, but has a population trajectory determined by a specified rate trajectory. It is also less flexible. All cohorts have the same base PTR matrix, there is a fixed allocation of births to states, and fertility and survivorship both change exponentially over time. To simplify the presentation, we again focus on a three-age, two living state model, but the approach is valid for any number of ages and states.

The MASM matrix that projects the time t − 1 population, at all ages, to time t, is once again denoted $\mathbf{A}_t$. That PPM is expressed as the product

$$\mathbf{A}_t = \wp_t\, \mathbf{B}_t\, \boldsymbol{\Phi}\, \mathbf{B}^{-1}{}_t\, \wp^{-1}{}_t \qquad (9.15)$$

Note that all time varying matrices have the same time index. Here, to define each of the matrices on the right hand side of equation (9.15), we start at the left and work inward.

MASM advancement matrix $\wp_t$ is defined by

$$\wp_t = \begin{bmatrix} \mathbf{I} & \mathbf{0} & \mathbf{0} \\ \mathbf{0} & \mathbf{P}^{\alpha+\gamma t} & \mathbf{0} \\ \mathbf{0} & \mathbf{0} & \mathbf{P}^{\alpha+\beta+t(\gamma+\delta)} \end{bmatrix} \qquad (9.16)$$

Base PTR advancement matrix $\mathbf{P}$ is a 2 × 2 matrix with known, constant elements, i.e. those of matrix $\boldsymbol{\Pi}$ in equation (9.7); parameter $\alpha$ reflects the initial (time 0) adjustment factor associated with advancement from age group 1 to age group 2; $\beta$ reflects the initial (time 0) adjustment factor associated with advancement from age group 2 to age group 3; $\gamma$ gives the rate of change over time in the adjustment factor from age group 1 to age group 2; and $\delta$ gives the rate of change in the adjustment factor from age group 2 to age group 3.

Diagonal birth cohort size matrix $\mathbf{B}_t$ is defined by

$$\mathbf{B}_t = \begin{bmatrix} \mathbf{I} & \mathbf{0} & \mathbf{0} \\ \mathbf{0} & \rho^{-1}\kappa^{-t}\mathbf{I} & \mathbf{0} \\ \mathbf{0} & \mathbf{0} & \rho^{-2}\kappa^{-2t}\mathbf{I} \end{bmatrix} \qquad (9.17)$$

where again $\mathbf{I}$ is the identity matrix and $\mathbf{0}$ a matrix of zeroes. Parameter $\rho$ reflects the growth rate of the (implicit) initial stable population and parameter $\kappa$ the rate of change over time in the stable growth implied by each $\mathbf{A}_t$. Note that $\mathbf{B}_t$ does not allocate births across states.

Matrix $\mathbf{\Phi}$ is a time invariant, block Leslie matrix analogous to the $\mathcal{F}$ matrix of equation (9.2), specifically

$$\mathbf{\Phi} = \begin{bmatrix} \mathbf{\Phi}_1 & \mathbf{\Phi}_2 & \mathbf{\Phi}_3 \\ \mathbf{I} & \mathbf{0} & \mathbf{0} \\ \mathbf{0} & \mathbf{I} & \mathbf{0} \end{bmatrix} \qquad (9.18)$$

where the age-state-specific MASM net maternity matrices are given by

$$\mathbf{\Phi}_i = \begin{bmatrix} \phi_{i11} & \phi_{i21} \\ \phi_{i12} & \phi_{i22} \end{bmatrix} \qquad (9.19)$$

with $\phi_{ijk}$ denoting the number of persons in age group 1, state k, at the end of the interval per person in age group i, state j, at the beginning of the interval. However, the MASM $\mathbf{\Phi}$ is not row stochastic. Instead, it satisfies the renewal condition in equation (9.22), as discussed below.

Using equations (9.15) through (9.19), the MASM PPM can be written

$$\mathbf{A}_t = \begin{bmatrix} \rho\kappa^t\,\mathbf{\Phi}_1 & \rho^2\kappa^{2t}\,\mathbf{\Phi}_2\,\mathbf{P}^{-\alpha-\gamma t} & \rho^3\kappa^{3t}\,\mathbf{\Phi}_3\,\mathbf{P}^{-\alpha-\beta-t(\gamma+\delta)} \\ \mathbf{P}^{\alpha+\gamma t} & \mathbf{0} & \mathbf{0} \\ \mathbf{0} & \mathbf{P}^{\beta+\gamma t} & \mathbf{0} \end{bmatrix} \qquad (9.20)$$

The MASM population vector at time t, $\mathbf{x}_t$, is given by

$$\mathbf{x}_t = \begin{bmatrix} \rho^t\kappa^{t(t+1)/2}\,\mathbf{g} \\ \rho^{t-1}\kappa^{t(t-1)/2}\,\mathbf{P}^{\alpha+\gamma t}\,\mathbf{g} \\ \rho^{t-2}\kappa^{(t-1)(t-2)/2}\,\mathbf{P}^{\alpha+\beta+\gamma(t-1)+\delta t}\,\mathbf{g} \end{bmatrix} \qquad (9.21)$$

where $\mathbf{g}$ is the constant birth allocation vector whose form is given in equation (9.12).

The projection relationship of equation (9.13), i.e. $\mathbf{x}_t = \mathbf{A}_t\,\mathbf{x}_{t-1}$, must hold. Under equations (9.20) and (9.21), that implies

$$\mathbf{g} = [\mathbf{\Phi}_1 + \kappa\,\mathbf{\Phi}_2\,\mathbf{P}^{-\gamma} + \kappa^3\,\mathbf{\Phi}_3\,\mathbf{P}^{-2\gamma-\delta}]\,\mathbf{g} \qquad (9.22)$$

Equation (9.22) is a renewal relationship analogous to equation (7.38) in the birth-death metastable model.

There are many ways to choose the elements of the $\mathbf{\Phi}_j$ so that equation (9.22) holds. One reasonable way to do so is to start with the elements of a row stochastic $\mathcal{F}$, and multiply row j of every $\mathbf{F}_i$ by an appropriate adjustment factor, say $k_j$. Let adjustment matrix $\mathbf{K}$ be given by

$$\mathbf{K} = \begin{bmatrix} k_1 & 0 \\ 0 & k_2 \end{bmatrix} \qquad (9.23)$$

with

$$\mathbf{\Phi}_i = \mathbf{K}\ \mathbf{F}_i \tag{9.24}$$

With n living states, equations (9.22)–(9.24) lead to n simultaneous equations that can be solved for the n unknown $k_j$.

To solve the 2 living state case, let $2 \times 2$ matrix $\mathbf{Y}$ with (i,j)th element y(i,j) be defined by

$$\mathbf{Y} = \mathbf{F}_1 + \kappa\ \mathbf{F}_2\ \mathbf{P}^{-\gamma} + \kappa^3\ \mathbf{F}_3\ \mathbf{P}^{-2\gamma-\delta} \tag{9.25}$$

Since $\mathbf{KY}$ is equal to the expression in brackets on the right hand side of equation (9.22), it follows from equations (9.22)–(9.25) that

$$k_1 = (1-g)/[(1-g)\ y(1,1) + g\ y(1,2)]; \quad k_2 = g/[(1-g)\ y(2,1) + g\ y(2,2)] \tag{9.26}$$

With the elements of $\mathbf{K}$ and thus the $\mathbf{\Phi}_i$ determined, the projection relationship holds and the MASM model is fully specified.

The form of equation (9.15) makes it possible to analytically determine product matrix $\mathbf{M}_{0,t} = \mathbf{A}_t\ \mathbf{A}_{t-1} \ldots \mathbf{A}_1$. Writing out that product, we find

$$\mathbf{M}_{0,t} = \rho^t \kappa^{t(t+1)/2}\ \wp_t\ \mathbf{B}_t\ \mathbf{S}^t\ \mathbf{B}^{-1}{}_0\ \wp^{-1}{}_0 \tag{9.27}$$

where

$$\mathbf{S} = \begin{bmatrix} \mathbf{\Phi}_1 & \kappa\ \mathbf{\Phi}_2\ \mathbf{P}^{-\gamma} & \kappa^2\ \mathbf{\Phi}_3\ \mathbf{P}^{-\gamma-\delta} \\ \mathbf{I} & \mathbf{0} & \mathbf{0} \\ \mathbf{0} & \kappa\ \mathbf{P}^{-\gamma} & \mathbf{0} \end{bmatrix} \tag{9.28}$$

Time invariant MASM synthesis matrix $\mathbf{S}$ has the same characteristic equation as matrix $\mathbf{KY}$, the bracketed expression on the right hand side of equation (9.22). Its dominant root, $\lambda_s$, is one (see Exercise 2), hence $\mathbf{S}$ does not affect the long term rate of population growth. When $\mathbf{S}^T$ becomes a rank one matrix, the long term effect of initial population size and composition is again simply that of a constant overall factor.

The MASM model can serve as a framework for analyzing patterned change in multi-age, multistate models. For example, one can consider the three living state marriage, divorce, and remarriage model of Subsection 8.2.4 in the context of continually rising transfer rates and declining birth cohort sizes. With $\rho = \kappa = 1$, uniform birth cohorts produce a wedge-period population that can facilitate the analysis of monotonically changing interstate transfer rates.

## 9.4 SUMMARY AND OVERVIEW

### 9.4.1 Chapter Summary

Dynamic multi-age, multistate models are potentially valuable demographic tools that go beyond conventional single state interval-by-interval projection. Two approaches are described that provide closed form solutions linking population size and age-state composition trajectories with generating sequences of population projection matrices. Hyperstable models allow any specified birth trajectory, with each birth cohort advancing according to its given PTR base matrix and set of age and period influences. Metastable models allow less flexibility, but give the age and state population trajectory produced by a set of changing rates of birth and cohort advancement. Further development of these models can enhance analyses that simultaneously examine changing rates of birth, death, and interstate movement.

### 9.4.2 Overview and Recapitulation

This book emphasizes the implications of changing demographic behavior. It focuses on occurrence/exposure rates of demographic events, exploits the logical closure of population dynamics, and stresses the close relationships between population stocks and flows.

*Chapter 1* discusses the classical fixed rate models of demography, both stationary and stable, and shows how they can be calculated using a General Algorithm. The decrement only life table describes the life course of a cohort. Multistate life tables extend that basic model to include movements between a broad range of living states. The stable population model adds fertility behavior and population renewal, while the multistate stable population can reflect the implications of given rates of fertility, mortality, and migration, the three core population processes.

*Chapter 2* focuses on convergence to the stable state, the conditions necessary for convergence, and the speed at which convergence takes place. *Chapter 3* considers population momentum, the change in population size that generally accompanies a change in vital rates. In modern demographic history, the demographic transition from high to low rates of birth and death has produced both substantial population growth and unprecedented population aging. *Chapter 4* examines how demographic rates influence population size and structure at the margin. Age-specific growth rates link age-specific behavioral change with overall demographic summary measures. Every population, at every time point, is always moving toward the stable population implied by its prevailing rates.

*Chapter 5* explores the relationships between period and cohort measures of mortality. When initial cohort size does not change over time, changing mortality creates a new perspective, the wedge-period. Average Cohort Life Expectancy (ACLE) provides the average longevity of all cohorts alive in a given year. Mortality dynamics can be analyzed using a shifting Gompertz mortality model. When death rates decline linearly over time, the model yields simple relationships for changes in longevity and population size and structure. *Chapter 6* looks at timing effects in fertility analyses, where changes in cohort timing are known to affect the level of period measures. Average Cohort Fertility (ACF) can measure the average fertility of cohorts reproducing in a given year, and shows the substantial effects timing factors had on 20th century American fertility. Extensions of the average cohort approach reveal considerable timing effects on past levels of nuptiality and divorce.

*Chapter 7* examines birth-death (multi-age) models with changing vital rates. If fertility varies according to an exponentiated mth degree polynomial, the birth trajectory follows an exponentiated polynomial of degree $m + 1$; if fertility fluctuates cyclically, so do births. Two approaches to finding the birth trajectory associated with a given sequence of vital rates are examined. The metastable approach yields a closed form solution for births when fertility is changing exponentially. Intrinsically dynamic models (IDMs) yield solutions for any pattern of fertility change. However, they require that the dominant roots of the population projection matrices be multiplicative over age/time intervals, and involve convergent infinite series. Hyperstable models, which assume a fixed proportional distribution of births by age of mother (i.e. by origin state), can yield sequences of Leslie matrices consistent with any given birth trajectory. Assuming that the age-specific fertility schedule changes proportionally at all ages can do so as well. In sum, a considerable number of tools are now available for analyzing dynamic multi-age models.

*Chapter 8* considers multistate models with changing rates. The Proportional Transfer Rate (PTR) approach simplifies cohort analyses when all transfer rates vary proportionally with age. The metastable, IDM, and hyperstable approaches extend to the multistate case, and yield noteworthy relationships. However, those approaches do not provide satisfactory estimates of transfer rates from population distributions. Two procedures that do, Iterative Proportional Fitting (IPF) and Relative State Attraction (RSA), give very similar results from very different premises. *Chapter 9* considers dynamic models with multiple ages and states. The study of those complex models is in its early stages, though both the hyperstable and metastable approaches can be brought to bear.

In efforts to model changing rates, it is necessary to come to grips with the complexity embodied in the seemingly simple differential equation

$$x' = \mu x \qquad (9.29)$$

where x represents a population, the prime indicates differentiation with respect to time, and $\mu$ represents demographic behavior. When those quantities are scalars, they indicate that constant demographic behavior implies an exponential population trajectory. With changing behavior and multiple ages or states, solutions to equation (9.29) are generally only calculable, not expressible analytically. However, metastable models, IDMs, and hyperstable models offer special cases where analytical solutions are possible. Other demographically plausible and analytically tractable dynamic models undoubtedly exist, and are worthy goals for future research.

Throughout, this book is concerned with change. We live in a rapidly globalizing world where many nations are experiencing profound demographic transformations, and where developing countries closely interact with post-industrial, post-Zero Population Growth societies. In that context, it is fitting for demographic analysis to move beyond fixed rates and structures to encompass processes of change and confront the challenges of dynamic multi-age and multi-state modeling.

## 9.5 EXERCISES

[See Appendix B for selected answers]

1. Given beginning and end of interval population distributions by age and state, and appropriate standard values, verify that the RSA approach of Section 8.6 can be used to find fertility and interstate transfer rates by proceeding one age group at a time.
2. Show that equation (9.22) implies that the determinant $|\mathbf{S}-\mathbf{I}|=0$, and therefore $\mathbf{S}$ has a dominant root of 1.
3. Algebraically verify
   a. the MASH projection relationship in equation (9.13)
   b. that, given equation (9.22), projection equation (9.13) holds in MASM models
4. Consider a 2 living state, 3 age group MASM model where changes over age occur at the same pace as changes over time (i.e. in equation (9.20), $\alpha=\beta$ and $\gamma=\delta$), and where $\rho=\kappa=1$.
   a. Verify that the synthesis matrix is given by

$$\mathbf{S}=\begin{bmatrix}\mathbf{\Phi}_1 & \mathbf{\Phi}_2\,\mathbf{P}^{-\gamma} & \mathbf{\Phi}_3\,\mathbf{P}^{-2\gamma}\\ \mathbf{I} & \mathbf{0} & \mathbf{0}\\ \mathbf{0} & \mathbf{P}^{-\gamma} & \mathbf{0}\end{bmatrix}$$

b. Confirm that the dominant right eigenvector of **S** can be written

$$\mathbf{u}_s = \begin{bmatrix} \mathbf{w} \\ \mathbf{w} \\ \mathbf{P}^{-\gamma}\ \mathbf{w} \end{bmatrix}$$

where **w** is given by

$$\mathbf{w} = \begin{bmatrix} 1 \\ g/(1-g) \end{bmatrix}$$

with g the fraction of births allocated to state 2. Show that in **S** matrices with additional age groups the jth vector element of $\mathbf{u}_s$ is $\mathbf{P}^{-\gamma(j-1)(j-2)/2}\ \mathbf{w}$.

c. Let $\gamma = 0.1$, $g = 0.3$, and

$$\mathbf{P} = \begin{bmatrix} .8 & .1 \\ .2 & .9 \end{bmatrix}$$

$$\mathbf{F}_1 = \begin{bmatrix} .25 & .05 \\ .1 & .2 \end{bmatrix}$$

$$\mathbf{F}_2 = \begin{bmatrix} .5 & .1 \\ .15 & .45 \end{bmatrix}$$

$$\mathbf{F}_3 = \begin{bmatrix} .08 & .02 \\ .03 & .07 \end{bmatrix}$$

Find the **K** matrix that yields $\boldsymbol{\Phi}_i$ values satisfying equation (9.22).

5. Consider the mixed MASH/MASM model where

$$\mathbf{A}_t - \wp_t\ \mathbf{B}_t\ \boldsymbol{\Phi}\ \mathbf{B}^{-1}{}_{t-1}\ \wp^{-1}{}_t$$

where $\wp_t$ and $\boldsymbol{\Phi}$ are defined as in equation (9.15), but

$$\mathbf{B}_t = \begin{bmatrix} x_t\mathbf{I} & \mathbf{0} & \mathbf{0} \\ \mathbf{0} & x_{t-1}\mathbf{I} & \mathbf{0} \\ \mathbf{0} & \mathbf{0} & x_{t-2}\mathbf{I} \end{bmatrix}$$

and

$$\mathbf{x}_t = \begin{bmatrix} x_t\ \mathbf{g} \\ x_{t-1}\ \mathbf{P}^{\alpha+\gamma t}\ \mathbf{g} \\ x_{t-2}\ \mathbf{P}^{\alpha+\beta+\gamma(t-1)+\delta t}\ \mathbf{g} \end{bmatrix}$$

The model has metastable change in **P**, but allows any arbitrary birth sequence. Verify that

a. the renewal condition analogous to equation (9.22) is

$$\mathbf{g} = [\mathbf{\Phi}_1 + \mathbf{\Phi}_2\,\mathbf{P}^{-\gamma} + \mathbf{\Phi}_3\,\mathbf{P}^{-2\gamma-\delta}]\,\mathbf{g}$$

b. the synthesis matrix is

$$\mathbf{S} = \begin{bmatrix} \mathbf{\Phi}_1 & \mathbf{\Phi}_2\,\mathbf{P}^{-\gamma} & \mathbf{\Phi}_3\,\mathbf{P}^{-\gamma-\delta} \\ \mathbf{1} & \mathbf{0} & \mathbf{0} \\ \mathbf{0} & \mathbf{P}^{-\gamma} & \mathbf{0} \end{bmatrix}$$

which has a dominant eigenvalue of 1.

c. the product matrix is

$$\mathbf{M}_{0,t} = \wp_t\,\mathbf{B}_t\,\mathbf{S}^t\,\mathbf{B}^{-1}{}_0\,\wp^{-1}{}_0$$

6. In the MASH model of equation (9.10), with **X** constant, show that the z(j,t) function can be seen as specifying age and period effects, while the relative sizes of the elements of PTR matrix **Π** can be seen as specifying noncollinear cohort effects.

# APPENDIX A

# MATRICES AND THEIR EIGENSTRUCTURE

This appendix provides the basic knowledge of matrix algebra needed to follow the material in the text. The mathematical level is kept as low as possible, even at the risk of some oversimplification. Those seeking a more thorough, though still abbreviated, discussion are referred to the appendix in Caswell (2001). The classic text on matrices is Gantmacher (1959). Some standard texts are Franklin (1968) and Horn and Johnson (1985). Seneta (1981) is a more advanced text on nonnegative matrices.

A matrix is an ordered, rectangular array of r rows and c columns. The (i,j)th element of a matrix is the element in row i and column j. Two matrices are equal when all of their corresponding elements are equal. Here we focus on square matrices with $\gamma$ rows and $\gamma$ columns. We also consider vectors, or ordered sequences. A column vector with $\gamma$ elements can be considered a $\gamma \times 1$ matrix, and a row vector with $\gamma$ elements can be considered a $1 \times \gamma$ matrix. A diagonal matrix has nonzero elements only on its main diagonal, i.e. the elements where $i = j$.

Let **A, B**, and **C** be $\gamma \times \gamma$ matrices whose (i,j)th elements are $a_{ij}$, $b_{ij}$, and $c_{ij}$ respectively. Addition involves summing corresponding elements. Thus $\mathbf{A} + \mathbf{B} = \mathbf{C}$ when $a_{ij} + b_{ij} = c_{ij}$ for all i and j. Multiplication by a scalar (ordinary number) involves multiplying each element of the matrix by that scalar. Multiplication of two vectors and/or matrices is more complicated. The product of two matrices is only defined when the number of rows in the first (leftmost) matrix is equal to the number of columns in the second. If $\mathbf{AB} = \mathbf{C}$, then

$$c_{ij} = \Sigma\, a_{ik}\, b_{kj} \qquad (A.1)$$

where the sum over k goes from 1 to $\gamma$. Thus the (i,j)th element of **AB** is found by summing the products obtained from multiplying corresponding elements in the ith row of **A** and the jth column of **B**.

Matrix algebra satisfies the Associative Law ($\mathbf{A} + \mathbf{B} = \mathbf{B} + \mathbf{A}$) and the Distributive Law ($\mathbf{A}[\mathbf{B} + \mathbf{C}] = \mathbf{AB} + \mathbf{AC}$). However matrix multiplication does *not* follow the Commutative Law (i.e. **AB** is not generally equal to **BA**). Matrix multiplication is commutative, however if both matrices are diagonal, that is both matrices have all of their off diagonal elements equal to zero.

Matrix division is accomplished by inversion. The inverse of matrix **A**, denoted $\mathbf{A}^{-1}$, is a matrix such that

$$\mathbf{A}^{-1}\,\mathbf{A} = \mathbf{A}\,\mathbf{A}^{-1} = \mathbf{I} \tag{A.2}$$

where **I**, the identity matrix, is a $\gamma \times \gamma$ diagonal matrix whose diagonal elements are 1. Multiplying a matrix by **I** leaves that matrix unchanged. Only square matrices with nonzero determinants can be inverted. The determinant is a unique scalar associated with a matrix, and matrices with a determinant of zero are termed "singular". Most of the demographic matrices we deal with are non-singular and can be inverted. The inverse of a diagonal matrix is a diagonal matrix whose ith diagonal element is the reciprocal of the ith diagonal element of the original matrix.

Eigenstructure refers to a unique decomposition possible with non-singular, square matrices. The eigenstructure of $\gamma \times \gamma$ population projection matrix **A** can be written

$$\mathbf{A} = \mathbf{U}\boldsymbol{\Lambda}\mathbf{V} \tag{A.3}$$

where **U** is a $\gamma \times \gamma$ matrix whose first row elements are equal to 1, $\Lambda$ is a $\gamma \times \gamma$ diagonal matrix, and **V** is the inverse of **U**.

Matrix $\boldsymbol{\Lambda}$, whose ith diagonal element is $\lambda_i$, is the matrix of eigenvalues (also called roots or characteristic values or latent values). The matrix is typically written with $\lambda_1$ being the largest root, $\lambda_2$ the second largest, and so on. The $\lambda_i$ are the solutions to the equation

$$|\mathbf{A} - \lambda\mathbf{I}| = 0 \tag{A.4}$$

where the vertical bars in equation (A.4) indicate the determinant. Equation (A.4), which is known as the characteristic equation of **A**, is a system of $\gamma$ equations in $\lambda$, giving rise to $\gamma$ roots. For the most part, we assume that the roots are distinct, as is generally the case in demography. When $\gamma > 2$, at least some roots are usually complex conjugates (i.e. expressions of the form $[\alpha + \beta i]$ and $[\alpha - \beta i]$, where $\alpha$ and $\beta$ are real numbers and $i$ is the square root of minus one). Demographically, the roots are growth rates. The dominant root, $\lambda_1$, is associated with the dominant or stable component of the population. From equation (1.44), $\lambda_1 = \exp(\mathrm{n}\ \mathrm{r}_1)$, where $\mathrm{r}_1$ is Lotka's r and n is the length of the interval. At stability, the dominant component is the entire population. The subordinate $\lambda_i$ are the growth rates of subordinate components, which by definition grow more slowly. The sum of the roots of a matrix is known as its trace, and is equal to the sum of the diagonal elements of the matrix.

Each column of **U** represents the relative age composition of one population component. The leftmost column vector, $\mathbf{u}_1$, represents the number of persons in each age group of the stable population relative to the number in the first age group. Succeeding columns of **U** represent the relative age compositions of subordinate components. The **u** vectors, the right eigenvectors of **A**, are found by solving the basic eigenstructure equation

$$\lambda\ \mathbf{u} = \mathbf{A}\ \mathbf{u} \tag{A.5}$$

which is identical in form to stable population projection equation (1.45). Equation (A.5) can be interpreted as indicating that projecting any right eigenvector ahead one interval is equivalent to multiplying that eigenvector by its eigenvalue. The eigenstructure renders the matrix multiplication of a vector equivalent to a scalar multiplication.

Matrix **V**, the inverse of **U**, is demographically meaningful in its own right. Each row of **V** is a left eigenvector, defined by the equation

$$\lambda\ \mathbf{v}' = \mathbf{v}'\mathbf{A} \tag{A.6}$$

where the prime (') indicates a transpose, i.e. that $\gamma \times 1$ column vector **v** becomes a $1 \times \gamma$ row vector. The first row of **V** represents the dominant left eigenvector and contains reproductive contributions. Specifically, the ith row element gives the contribution of each person in the ith category of the population being projected to the first category of the ultimate stable population. In the Leslie (birth-death) matrix, $v_{1j}$, the jth element of the dominant left eigenvector, is a contribution to future births. It gives the contribution of each person in the jth age group to the stable equivalent number of births, that is the number of births in the ultimate stable population discounted for growth to the initial time point. Eigenvectors are defined only up to a scalar factor. Conventionally, they are scaled so that $\mathbf{v}'\mathbf{u} = 1$.

For the most part, finding the eigenstructure of a matrix is best left to a mathematical computer package (such as MATLAB, MAPLE, or MATHEMATICA). The eigenstructure of a $2 \times 2$ matrix is fairly simple, however, and worth examining. Let our matrix be population projection matrix **A**. Then we have

$$\mathbf{A} = \begin{bmatrix} a & b \\ c & d \end{bmatrix} \tag{A.7}$$

and

$$\mathbf{\Lambda} = \begin{bmatrix} \lambda_1 & 0 \\ 0 & \lambda_2 \end{bmatrix} \tag{A.8}$$

where $\lambda_{1,2} = \{a + d \pm [(a - d)^2 + 4bc]^{.5}\}/2$. The eigenvector matrices are

$$\mathbf{U} = \begin{bmatrix} 1 & 1 \\ c/(\lambda_1 - d) & c/(\lambda_2 - d) \end{bmatrix} \qquad (A.9)$$

and

$$\mathbf{V} = \begin{bmatrix} \lambda_1 - d & -(\lambda_1 - d)(\lambda_2 - d)/c \\ -(\lambda_2 - d) & (\lambda_1 - d)(\lambda_2 - d)/c \end{bmatrix} \{1/(\lambda_1 - \lambda_2)\} \qquad (A.10)$$

The determinant of **A** is $(ad - bc)$, and the trace is $(a + d)$. In the $2 \times 2$ Leslie matrix, $d = 0$, and the eigenvector matrix elements simplify.

The use of eigenstructure can simplify functional transformations of matrices. Consider the rate matrix $\boldsymbol{\mu} = \mathbf{URV}$. The exponential of that rate matrix is population projection matrix **A**, which, for a unit projection interval, can be written

$$\mathbf{A} = \mathbf{exp(URV)} = \mathbf{U}\ \mathbf{exp(R)}\ \mathbf{V} \qquad (A.11)$$

where $\exp(\mathbf{R})$ is defined as $\boldsymbol{\Lambda}$. However, the Leslie matrix is a PPM that does not have a well defined underlying rate matrix because its states are based on age. Employing eigenstructure is useful in raising a matrix to a power. Since $\mathbf{VU} = \mathbf{I}$, the matrix $\mathbf{A}^2 = \mathbf{AA}$ can be written

$$\mathbf{A}^2 = (\mathbf{U\Lambda V})(\mathbf{U\Lambda V}) = \mathbf{U\Lambda}^2\mathbf{V}$$

and in general

$$\mathbf{A}^t = \mathbf{U\Lambda}^t\mathbf{V} \qquad (A.12)$$

# APPENDIX B

# SELECTED ANSWERS TO EXERCISES

## CHAPTER 1: POPULATION MODELS WITH CONSTANT RATES

1. M must exceed 1 as there is one death for less than a person-year of life. The value of $\mu$ must become infinite to drive survivorship to zero.
2. $EPL = [T(15) - T(65)]/\ell(15)$
3. $e(90) = a(90, \infty) = 1/M(90, \infty)$
4. a. $\mu = 1/(\omega - x)$ b. $e(0) = \omega/2$
6. It follows from $\mu(x) = \mu^{(i)}(x) + \mu^{(-i)}(x)$
7. $D\ T(65)/[T(20) - T(65)]$
9. Use equations (1.32) and (1.33) and $b = r + d$ to iteratively solve for r. The relationship between the probabilities of survival and the crude death rate imply r.
12. $a_0 = 17/35$, $a_1 = 12/35$, $a_2 = -3/35$, and the smoothed value is 18.286.
14. Scale the observed population so that the number in the first age group is 1. Then the number in the ith age group, $x_i = p_i\lambda^{i-1} = s_1 s_2 \ldots s_{i-1}\lambda^{i-1}$. With $\gamma$ age groups, those relationships lead to $\gamma - 1$ equations of the form $s_i\lambda^{-1} = x_{i+1}/x_i^2$. Select the largest ratio of x's, and set the associated $s_i = 1$. Then use the $\gamma - 1$ equations to find $\lambda$ and the remaining s values. The solution is not unique as any nonzero scalar multiple will satisfy the equations.

## CHAPTER 2: CONVERGENCE TO STABILITY

2. a. It is reducible. b. The first row elements are (89 0 55 0).
3. a. Cycle length 2, roots 1 and −1. b. Cycle length 3, roots are the cube roots of 1, i.e. 1, $-\frac{1}{2}[1 \pm i(3)^{1/2}]$
4. The t-th power of the PPM is

$$\begin{bmatrix} e^{\alpha t} & 0 \\ \frac{\mu_{AB}(e^{\alpha t} - e^{\beta t})}{(\alpha - \beta)} & e^{\beta t} \end{bmatrix}$$

8. It follows directly from $e^x \approx 1 + x$.

10. At time 15, let the number of births, adjusted for mortality, be the number in the third age group of the desired stable population at time 45. At time 30, let the number of births, adjusted for mortality, be the number in the second age group of the desired stable population at time 45. At time 45, let the number of births be that of the desired stable population, and the stable composition is in place through age 45.

## CHAPTER 3: POPULATION MOMENTUM

4c. The initial population under age 30 is 1.8333. In a the ultimate population under 30 is 1.82689, and in b it is 1.83642. Equation (2.27) gives Q = 0.91667, or an ultimate population under 30 of 1.8333.

5. The projected population can grow to infinity. One way to demonstrate that is to project by 3 PPMs that are repeated indefinitely. Let the first of those PPMs be

$$\begin{bmatrix} 0 & 0 & 4 \\ .5 & 0 & 0 \\ 0 & .5 & 0 \end{bmatrix}$$

and let the second and third PPMs be

$$\begin{bmatrix} 1 & 0 & 0 \\ 1 & 0 & 0 \\ 0 & 1 & 0 \end{bmatrix}$$

Then after the first cycle of projection by the 3 PPMs, there are 4 persons at each age. After the second cycle there are $4^2$, and after the nth there are $4^n$.

6. No, because it would produce negative elements in **B**, e.g. element (3,3).

## CHAPTER 4: DEMOGRAPHIC CHANGE AT THE MARGIN

1. From equation (4.2), $p^*(x) = p(x)e^{-cx}$. The age composition does not change because $r^* = r - c$.
2. It follows because equation (4.9) yields $\delta p(x)/p(x) = \ln[p(x)^k] \approx p(x)^k - 1$, which is the result given by equation (4.4).
3. The maximum of 5/8 occurs when $r = 0$.
5. Integrate the series for $A^*$ and relate to equation (4.28).

6–9. See Schoen and Kim (1992)

10. The relationship holds as $q_{11} = 1/[1 + x(1 - a)]$, $q_{12} = x(1 - a)/[1 + x(1 - a)]$, $q_{21} = 1/[1 - x]$, and $q_{22} = -x/[1 - x]$ while $\Omega_{11} = [1 + x(1 - a)]/$

$(2 - a)$, $\Omega_{12} = [1 + x(1 - a)]/[x(2 - a)]$, $\Omega_{21} = (1 - x)(1 - a)/[2 - a]$, and $\Omega_{22} = -[1 - x]/[x(2 - a)]$, remembering that $\Omega_{21}$ is the (1,2) element of matrix $\boldsymbol{\Omega}$.

## CHAPTER 5: LONGEVITY AND DYNAMIC MORTALITY

1. Expand the square in $\int (x - e(0))^2 d(x)\, dx$, and use integration by parts to simplify the integral with $x^2$.
2. In the wedge-period context, $Q = 1$ and $b = 1/CAL(t)$.
8. With $\ell_{LT}(x)$ representing life table survivorship to age x and $d_{LT}(x)$ representing the number of deaths between ages x and $x + n$ in that life table, we seek the age where $d_{LT}(x)$ is a maximum. Differentiation yields

$$d\, d_{LT}(x)/dx = d_{LT}(x)[d \ln \mu(x)/dx - \mu(x)]$$

   For $d_{LT}(x) \neq 0$, setting the derivative equal to zero gives $d \ln \mu(x)/dx = \mu(x)$.
10. a. 1.8189 b. 1.9095 c. 2 d. 2.4513

## CHAPTER 6: TIMING EFFECTS ON FERTILITY, MARRIAGE, AND DIVORCE

5. Assume that the time t composition of the model population is that of the stable population implied by time t rates. With Lotka's r less than 0 and an absence of mortality below the highest age of reproduction ($\beta$), the number of persons in the stable population increases with age through $\beta$. A shift in the fertility curve to higher ages would thus increase the number of births, though the number of births in that stable population is decreasing over time. That contradiction indicates that the number of persons in the dynamic population must rise more slowly with age, and thus the dynamic population declines at a rate that is less negative than the stable rate.
6. The population does not grow but adopts a constant birth trajectory.

## CHAPTER 7: DYNAMIC BIRTH-DEATH (MULTI-AGE) MODELS

3b. That is usually a reasonable approximation.

5. It follows from equating proportions, as in equation (4.24). Here

$$\exp[-rB_4]p(B_4)\ f(B_4)/1 = \exp[-B_4(r + s) + hB_4^2/2]p(B_4)\ f(B_4)/1$$

8. For Pattern 2, $\mu_A = 22^2/_3$, $\mu_B = 27$, $Q = 0.83951$, and the initial and final numbers under age 30 are 1.66667 and 1.67901. For Pattern 4, $\mu_A = 24^1/_3$, $\mu_B = 30$, $Q = 0.81111$, and the initial and final numbers under age 30 are 1.66667 and 1.62222.

12c. The value of $\lambda_s$ is constant and cannot be made equal to the $k^{-j(j-1)/2}$, j=1,n.

## CHAPTER 8: DYNAMIC MULTISTABLE MODELS

3. The sum of the $y_j(x)$ as j goes from 0 to $\infty$ is 1 because it is equal to $\exp[-Z]$ times a power series that sums to $\exp[Z]$. The maximum of $y_j(x)$ is found by setting its derivative with respect to Z equal to zero.
4. Setting $dy_m/dZ = 0$ yields the relationship $\exp[-Z(\mu_{mv} + \mu_{vm} - \mu_{sm})] = (\mu_{sm} - \mu_{vm})/\mu_{mv}$.
5. Each PPM $\mathbf{A_t}$ is rank 1, i.e. row and column proportional. As a result, $\mathbf{A_t} = \lambda_t\, \mathbf{u_t}\, \mathbf{v_t}'$, where $\mathbf{v_t}' = (1 - u_t v_t, v_t)$ and $v_t$ is a scalar reproductive value. It follows that $\mathbf{v_t}'\mathbf{u_{t-1}}$ is a scalar while $\mathbf{v_t}'\mathbf{u_t} = 1$. PPMs of the form of $\mathbf{A_t}$ represent a readily soluble dynamic model, but the form of $\mathbf{A_t}$ is quite unusual for a demographic PPM.

9c. Because equation (8.31) is for a continuous model and equation (8.39) for a discrete model, they are not strictly comparable. However, using the linear approximation $e^{-x} \approx 1 - x$ in part a yields the solution in part b.

## CHAPTER 9: DYNAMIC MODELS WITH MULTIPLE AGES AND STATES

4c. $k_1 = 1.094074$; $k_2 = 0.738281$

# APPENDIX C

# INDEX OF THE PRINCIPAL SYMBOLS USED

| Symbol | Brief Description | Introduced in Equation(s) |
|---|---|---|
| **Boldface** | a matrix (or vector) | (1.42) |
| (x, n) | following a function, age group x to x+n | (1.2) |
| a (x, n) | number of years lived in an interval by persons dying in the interval (Chiang's a) | (1.16) |
| A | mean age at childbearing | (7.5) |
| $A^*$ | stable population mean age at childbearing | (1.37) |
| $A_{M1}$ | stable population mean age at first marriage | (4.27) |
| $A_N$ | observed population mean age at childbearing | after (3.7) |
| ACF | Average Cohort Fertility | (6.9) |
| ACLE | Average Cohort Life Expectancy | (5.1) |
| **A** | population projection matrix (PPM) | (1.42) |
| b | intrinsic birth rate | (1.33) |
| b | Gompertz parameter | (5.2) |
| $b_i$ | projection matrix first row fertility element | (1.41) |
| B (t) | number of births | (1.27) |
| BR | crude birth rate of a data population | (3.5) |
| **B** | block Leslie age-state PPM | (3.24) |
| c (x) | proportion of the population at age x last birthday | (1.35) |
| $c_i(x)$ | proportion of the population attaining age x in state i | (1.79) |
| $c_N(x, t)$ | population proportion at age x, time t | (4.35) |
| cov (X,Y) | covariance between variables X and Y | (4.41) |
| $C_L(0, G)$ | fraction of life table population between ages 0 and G | (3.7) |
| $C_N(0, G)$ | fraction of observed population between ages 0 and G | (3.7) |

*(Continued)*

*(Continued)*

| Symbol | Brief Description | Introduced in Equation(s) |
|---|---|---|
| CAL (t) | size of wedge-period population | (5.1) |
| d (x,n) | life table decrement | (1.7) |
| $d_{ij}(x, n)$ | number of decrements from state i to state j at ages (x,n) | (1.46) |
| **d (x,n)** | multistate matrix of decrements | (1.55) |
| D (x,n) | a data event or decrement | (1.1) |
| e (x) | an expectation of life | (1.11) |
| e (x,t) | life expectancy at age x under rates of time t | (5.12) |
| $e_c(0, t)$ | life expectancy of cohort born year t | (5.1) |
| $e_i(x)$ | population based life expectancy in state i | (1.71) |
| f (x) [f(x,n)] | fertility rate at age x [over (x,n) interval] | (1.27) |
| $F_{30}$ | ratio of initial to ultimate populations under age 30 | (3.14) |
| F (t) | shift in $\nu$ from 0 to t | (5.6) |
| **F** | base Leslie matrix | (7.25) |
| $\mathscr{F}$ | MASH stationary block Leslie matrix | (9.1) |
| g(x) | an age graded function | (4.13), (4.37) |
| g(t) | number of births at time t | before (7.1) |
| $\mathbf{g_t}$ | MASH/MASM birth allocation vector | (9.12) |
| G | a population summary measure | (4.13), (4.37) |
| $\mathbf{G_t}$ | diagonal matrix of births at (and before) time t | (7.94) |
| $G_L$ | cohort proportion with characteristic g | (4.16) |
| $G_P$ | stable population proportion with characteristic g | (4.15) |
| h* | ultimate level of the force of convergence | (2.15) |
| h (t) | force of convergence at time t | (2.14) |
| H | entropy | (2.17), (5.24) |
| *i* | square root of $-1$ | (2.5) |
| **I** | identity matrix | (A.2) |
| J (t) | shift in force of mortality from 0 to t | (5.10) |
| k | metastable growth factor | (7.23) |
| $k_j(x, u)$ | relative state attraction factor | (8.74) |
| K (t) | Kullback distance at time t | (2.11) |
| ln | natural logarithm | above (1.5) |
| $\ell(x)$ | number of survivors to exact age x | (1.3) |
| $\ell_c(x,\ t-x)$ | survivors to exact age x in cohort born time t-x | (5.1) |
| $\ell_i(x)$ | number of survivors to exact age x in state i | (1.46) |
| $\ell_{LT}(x, t)$ | survivors to exact age x in life table for time t | (5.11) |
| $\boldsymbol{\ell}(\mathbf{x}+\mathbf{n})$ | multistate survivorship matrix at age $x+n$ | (1.54) |
| L (x,n) | life table population/number of person-years at ages (x,n) | (1.9) |
| $L_i(x, n)$ | life table population in state i at ages (x,n) | (1.52) |

| | | |
|---|---|---|
| **L (x,n)** | multistate person-year matrix | (1.61) |
| **L (t)** | eigenvalue matrix of multistate PPM | (8.52), (8.54) |
| m (x,n) | a model occurrence/exposure rate | (1.10) |
| $m_{ij}$ (x, n) | model rate of transfer from state i to state j at ages (x,n) | (1.50) |
| **m(x,n)** | multistate matrix of model rates | (1.60) |
| MAC | Mean Age at Childbearing | (6.3) |
| M (x,n) | a data occurrence/exposure rate | (1.1) |
| $M_{ij}$(x, n) | data rate of transfer from state i to state j at ages (x,n) | (1.50) |
| M (t) | modal age at death at time t | (5.37) |
| M1 | stable population number of first marriages | (4.27) |
| **M (x,n)** | multistate data rate matrix | (1.59) |
| $\mathbf{M}_{0,t}$ | product matrix that projects from time 0 to time t | (7.26) |
| n (x) | stable population number at age x | (4.17) |
| N (x,t) | number of persons age x at time t | (2.7) |
| $N_{jt}$ | number of persons in age group j at time t | (4.54) |
| $N_{ijt}$ | number in the ith component, jth age group, at time t | (4.55) |
| NRR ($R_0$) | Net Reproduction Rate | (1.30) |
| p (x) | probability of surviving from birth to exact age x | (1.27) |
| $p_i$ | probability of surviving from age group 1 to age group i | (1.43) |
| $p_i$(x, n) | multistate p-probability of survival | (1.47) |
| P (x,n) | a data population between the ages of x and x+n | (1.1) |
| $\mathbf{P}_{jt}$ | MASH/MASM cohort advancement matrix | (9.6) |
| $\mathbf{P}(\mathbf{x}, \mathbf{w})$, $\mathbf{P}_{0,t}$ | product of population projection matrices | (8.5), (8.25) |
| $\wp_t$ | MASH/MASM survivorship matrix | (9.1) |
| PEM* | adjusted proportion ever marrying | (6.20) |
| PMED | proportion of marriages ending in divorce | (6.21) |
| PPM | population projection matrix | (1.41), (2.1) |
| q (x,n) | life table probability of decrement | (1.8) |
| q (x,t) | proportional contribution of the population age x at time t to stable equivalent births Q(t) | (2.9) |
| $q_{ijt}$ | contribution of the jth age group, ith component to the time t stable equivalent $Q_{it}$ | (4.63) |
| $\mathbf{q}_t$ | matrix of $q_{ijt}$ | (4.66) |
| Q | stable equivalent number [often of births] | (1.28), (2.4) |
| $Q_{it}$ | stable equivalent of component i at time t | (4.55) |
| r | Lotka's intrinsic stable rate of growth | (1.29) |
| r (x,t) | growth of population age x at time t | (4.18) |
| $r_o$ | change in the timing of fertility of order o | (6.3) |

*(Continued)*

*(Continued)*

| Symbol | Brief Description | Introduced in Equation(s) |
|---|---|---|
| $r_{jt}$ | growth rate of the jth age group between times $t-1$ and t | (4.54) |
| $\mathbf{r_t}$ | vector of age-specific growth rates | (4.62) |
| $\check{r}_j(t)$ | jth eigenvalue of latent rate matrix $\boldsymbol{\nu}$ | after (8.54) |
| $R_0$ | Net Reproduction Rate (NRR) | (1.30) |
| $R_i$ | ith moment of a distribution | (1.30) |
| R (t) | NRR at time t | (7.3) |
| **R** | eigenvalue matrix of $\mu$ | (2.38), (A.11) |
| $R_B(t)$ | rate of increase in births at time t | (4.33) |
| $R_Z$ | rate of increase in Z | (4.40) |
| $R_{Zg}$ | rate of increase in Z times g | (4.40) |
| $\check{\mathbf{R}}(\mathbf{t})$ | Diagonal matrix of eigenvalues of latent rate matrix $\boldsymbol{\nu}(\mathbf{t})$ | (8.54) |
| s | subordinate eigenvalue of $\boldsymbol{\mu}$ | (2.33) |
| $s_i$ | projection matrix survivorship element | (1.41) |
| s (x) | stable reproductive value density distribution | (2.10) |
| S (t) | offset that shifts $\nu$ to $\mu$ | (5.8) |
| S (x,t) | stable equivalent population age x at time t | (2.12) |
| **S (x,n)** | matrix of survivorship proportions | (1.69) |
| **S** | metastable synthesis matrix | (7.28) |
| T(x) | life table population over age x | (1.11) |
| $T^*(0)$ | total number of persons in a stable population | (1.32) |
| Ŧ | Lotka's length of generation | (1.38) |
| TI | Timing Index | (6.8), (6.18) |
| TFR | Total Fertility Rate | (6.1) |
| $TFR^*$ | Bongaarts-Feeney adjusted TFR | (6.5) |
| $u_{ij}$ | jth element of ith right eigenvector | (4.55) |
| **u** | dominant right eigenvector of **A** | (1.45), (A.5) |
| **U** | matrix of right eigenvectors of **A** | (2.1), (A.3) |
| $V_L$ | stationary population mean age of reproductivity | (3.8) |
| v (x) | reproductive contribution function | (2.8) |
| $v_{ij}$ | jth element of ith left eigenvector | (4.56) |
| **v'** | dominant left (row) eigenvector of **A** | (2.3), (A.6) |
| **V** | matrix of left eigenvectors of **A** | (2.1), (A.3) |
| **W (t)** | right eigenvector matrix of latent rate matrix $\boldsymbol{\nu}(\mathbf{t})$ | (8.52) |
| $x_{jt}$ | jth element of the time t population vector | (2.25) |
| $\mathbf{x_t}$ | population vector, time t | (1.42) |
| $\mathbf{X_t}$ | MASH birth matrix | (9.1) |
| **X (t)** | fundamental matrix of multistate solution vectors | (8.51) |
| $X_Z$ | crossover point for stable and stationary Z densities | (4.26) |

| | | |
|---|---|---|
| $y_j$ | number of persons in state j of a multistate population | (8.6) |
| $Y_j(x)$ | number of person-years lived in state j at age x | (8.9) |
| $Y_g$ | mean age of the population with characteristic g | (4.14) |
| $Y_{GL}$ | stationary population mean age with characteristic g | (4.15) |
| $Y_{L:}$ | stationary population mean age | (3.8) |
| $Y_N$ | mean age of observed population | (4.44) |
| $Y_r$ | mean age of a stable population | (1.36), (3.10) |
| $Y_\delta$ | population mean age at death | Ch.4/Ex.6,(5.16) |
| z (x,n) | level of rates factor | (8.2) |
| Z (x) | sum of level factors z (x,n) up to age x | (8.7) |
| Z (x,t) | a demographic variable that changes according to r (x,t) | (4.37) |
| $\beta(x, t)$ | Fraction of experience of the cohort born year t-x at age x | (6.7), (6.17) |
| $\gamma$ | dimension of a PPM | (1.41) |
| $\gamma(t)$ | gap between period and cohort life expectancy at time t | (5.33) |
| $\gamma^E$ | Euler's constant, approx. 0.577215 | (5.31) |
| $\delta$ | absorbing or "dead" state | (2.28) |
| $\delta$ | change operator | (4.9) |
| $\in_{ij}$ (x) | status based life expectancy | (1.72) |
| $\kappa_i$ | stable population mean age at childbearing, component i | (4.57) |
| $\lambda$ | stable population growth parameter, dominant root of **A** | (1.44) |
| $\lambda_f$ | forward lag at time t | (5.35) |
| $\lambda_i$ | ith eigenvector of a PPM | (4.55) |
| $\lambda_s$ | dominant eigenvalue of synthesis matrix S | (7.29) |
| $\boldsymbol{\lambda}$ | vector of component-specific eigenvalues | (4.62) |
| $\boldsymbol{\Lambda}$ | diagonal matrix of eigenvalues of **A** | (2.1), (A.3) |
| $\mu$ | mean age of net maternity/childbearing | (1.31), (7.63) |
| $\mu(x)$ | force of mortality (or other decrement) | (1.3) |
| $\mu_{ij}(x)$ | force of decrement from state i to state j at age x | (1.51) |
| $\mu_3$ | third cumulant of the cohort birth distribution | (4.28) |
| $\nu$ (x, t) | proportional change over age | (4.21), (5.3) |
| $\boldsymbol{\nu}$ **(t)** | matrix of latent rates | (8.58) |
| $\pi_{ij}(x, n)$ | multistate $\pi$-probability of survival | (1.49) |
| $\boldsymbol{\Pi}(\mathbf{x}, \mathbf{n})$ | multistate matrix of $\pi$-probabilities | (1.57) |
| $\boldsymbol{\Pi}_t$ | MASH base PTR matrix (Cyrillic P) | (9.7) |
| $\rho$ | multistate model rate of natural increase | (3.21) |
| $\sigma^2$ | variance of cohort net maternity | (1.31) |

*(Continued)*

*(Continued)*

| Symbol | Brief Description | Introduced in Equation(s) |
|---|---|---|
| $\sigma_Y^2$ | variance of stationary population age distribution | (3.10) |
| $\phi(x, t)$ | net maternity rate at age x and time t | (7.1) |
| $\mathbf{\Phi}$ | MASM block Leslie matrix | (9.15) |
| $\Omega$ | population momentum | (3.1) |
| $\Omega(x, t)$ | momentum of age x at time t | (4.49) |
| $\Omega_{ijt}$ | momentum of age group j, component i, at time t | (4.59) |
| $\mathbf{\Omega_t}$ | matrix of $\Omega_{ijt}$ | (4.59) |
| $\infty$ | infinity | (1.20) |

# REFERENCES


Alho, J.M. and Spencer, B.D. 2005. *Statistical Demography and Forecasting*. New York: Springer.

Alligood, K.T., Sauer, T.D. and Yorke, J.A. 1996. *Chaos: An Introduction to Dynamical Systems*. New York: Springer.

Arthur, W.B. 1984. The analysis of linkages in demographic theory. *Demography* 21: 109–28.

Arthur, W.B. and Vaupel, J.W. 1984. Some general relationships in population dynamics. *Population Index* 50: 214–26.

Batty, M. and Mackie, S. 1972. The calibration of gravity, entropy and related models of spatial interaction. *Environment and Planning* 4: 205–33.

Bennett, N.G. and Horiuchi, S. 1981. Estimating the completeness of death registration in a closed population. *Population Index* 47: 207–21.

Bernardelli, H. 1941. Population waves. *Journal of the Burma Research Society* 31: 1–18.

Birdsall, N. and Sinding, S.W. 2001. How and why population matters: New findings, new issues. In N. Birdsall, A.C. Kelley, and S.W. Sinding (eds) *Population Matters*. Oxford: Oxford University Press.

Bishop, Y.M.M., Fienberg, S.E., and Holland, P.E. 1975. *Discrete Multivariate Analysis: Theory and Practice.* Cambridge MA: MIT Press.

Bongaarts, J. 1998. Demographic consequences of declining fertility. *Science* 282(16): 419–20.

Bongaarts, J. 1999. The fertility impact of changes in the timing of childbearing in the developing world. *Population Studies* 53: 277–89.

Bongaarts, J. 2002. The end of the fertility transition in the developed world. *Population and Development Review* 28: 419–43, 589.

Bongaarts, J. 2005. Long range trends in adult mortality: Models and projection methods. *Demography* 42: 23–49.

Bongaarts, J. and Bulatao, R.A. 1999. Completing the demographic transition. *Population and Development Review* 25: 515–29.

Bongaarts, J. and Bulatao, R.A. (eds). 2000. *Beyond Six Billion: Forecasting the World's Population*. Washington DC: National Academy Press.

Bongaarts, J. and Feeney, G. 1998. On the quantum and tempo of fertility. *Population and Development Review* 24: 271–91.

Bongaarts, J. and Feeney, G. 2002. How long do we live? *Population and Development Review* 28: 13–29.

Boyce, W.E. and DiPrima, R.C. 1977. *Elementary Differential Equations and Boundary Value Problems* (3d ed). New York: Wiley.

Brouard, N. 1986. Structure et dynamique des populations. La pyramide des années à vivre, aspects nationaux et exemples régionaux. *Espaces, Populations, Sociétés* 2: 157–68.

Butz, W.P. and Ward, M.P. 1979. Will U.S. fertility remain low? A new economic interpretation. *Population and Development Review* 5: 663–88.

Canudas-Romo, V. 2005. The modal age at death and the shifting mortality hypothesis. Unpublished manuscript, Department of Demography, University of California, Berkeley.

Canudas-Romo, V. and Schoen, R. 2005. Age-specific contributions to changes in the period and cohort life expectancy. *Demographic Research* 13: 63–82.

Caswell, H. 2001. *Matrix Population Models* (2d ed). Sunderland MA: Sinauer.

Cerone, P. 1996. On the effects of the generalized renewal integral equation model of population dynamics. *Genus* 52: 53–70.

Chiang, C.L. 1964. A stochastic model of competing risks of illness and competing risks of death. In J. Gurland (Ed), *Stochastic Models in Medicine and Biology*. Madison: University of Wisconsin Press.

Chiang, C.L. 1968. *Introduction to Stochastic Processes in Biostatistics*. New York: Wiley.

Chilton, R. and Poet, R. 1973. An entropy maximizing approach to the recovery of detailed migration patterns from aggregate census data. *Environment and Planning* 5: 135–146.

Coale, A.J. 1972. *The Growth and Structure of Human Populations*. Princeton NJ: Princeton University Press.

Coale, A.J. and Demeny, P. 1966. *Regional Model Life Tables and Stable Populations*. Princeton NJ: Princeton University Press.

Coale, A.J. and Demeny, P. 1983. *Regional Model Life Tables and Stable Populations* (2d ed). New York: Academic Press.

Coale, A.J. and Trussell, T.J. 1974. Model fertility schedules: Variations in the age-structure of childbearing in human populations. *Population Index* 40: 185–258.

Cohen, J.C. 1982. Multiregional age-structured populations with changing rates: Weak and stochastic ergodic theorems. Pp. 477–503 in K.C. Land and A. Rogers (eds), *Multidimensional Mathematical Demography*. New York: Academic Press.

Deming, W.E. and Stephan, F.E. 1940. On a least squares adjustment of a sampled frequency table when the expected marginal totals are known. *Annals of Mathematical Statistics* 11: 427–44.

DeSantis, G. 2003. The demography of an equitable and stable intergenerational transfer system. *Population-E* 58: 587–622.

DuPasquier, L.G. 1912. Mathematische theorie der invaliditaetsversicherung. *Mitteilungen der Vereinigung Schweitzerischer Versicherungsmathematiker* 7: 1–7.

Easterlin, R.A. 1980. *Birth and Fortune*. New York: Basic Books.

Elo, I.T. and Preston, S.H. 1992. Effects of early-life conditions on adult mortality: A review. *Population Index* 58: 186–212.

Espenshade, T.J., Bouvier, L.F., and Arthur, W.B. 1982. Immigration and the stable population model. *Demography* 19: 125–33.

Euler, L. 1760 (orig.). A general investigation into the mortality and multiplication of the human species. Translated by N. Keyfitz and B. Keyfitz in *Theoretical Population Biology* 1 (1970): 307–14.

Feeney, G.M. 1970. Stable by region age distributions. *Demography* 7: 341–48.

Feichtinger, G. and Vogelsang, H. 1978. *Pseudostabile Bevolkerungen: populationsdynamik bei gleichmassig sinkender Fertilitat*. Schriftenreihe des Instituts fur Demographie der Osterreichischen Akademie der Wissenschaften, Vol 4, Vienna.

Feller, W. 1941. On the integral equation of renewal theory. *Annals of Mathematical Statistics* 12: 243–67.

Fix, E. and Neyman, J. 1951. A simple stochastic model of recovery, relapse, death, and loss of patients. *Human Biology* 23: 205–41.

Ford L.R. 1955. *Differential Equations*. New York: McGraw-Hill.

Franklin, J.N. 1968. *Matrix Theory*. Englewood Cliffs NJ: Prentice-Hall.

Frauenthal, J.C. 1975. Birth trajectory under changing fertility conditions. *Demography* 12: 447–54.

Frauenthal, J.C. and Swick, K.E. 1983. Limit cycle oscillations of the human population. *Demography* 20: 285–98.

Frejka, T. and Calot, G. 2001. Cohort reproductive patterns in low-fertility countries. *Population and Development Review* 27: 103–32.

Furness, K.P. 1965. Time function iteration. *Traffic Engineering and Control* 7: 458–60.

Gantmacher, F.R. 1959. *Matrix Theory*. New York: Chelsea.

Goldstein, J.R. 2002. Population momentum for gradual demographic transitions. *Demography* 39: 65–73.

Goldstein, J. 2004. Late but not never: The tempo and quantum of first marriage in France. Unpublished manuscript, Office of population Research, Princeton University.

Goldstein, J.R. 2006. Found in translation? A cohort perspective on tempo-adjusted life expectancy. *Demographic Research* 14: 71–84.

Goldstein, J.R. and Stecklov, G. 2002. Long-range population projections made simple. *Population and Development Review* 28: 121–41.

Goldstein, J.R. and Wachter, K.W. 2006. Relationships between period and cohort life expectancy: Gaps and lags. *Population Studies*, 60: 257–269.

Graunt, J. 1662 (orig). *Natural and Political Observations...Made Upon the Bills of Mortality*. Baltimore: Johns Hopkins Press, 1939.

Guillot, M. 2003. The cross-sectional average length of life (CAL): A cross-sectional mortality measure that reflects the experience of cohorts. *Population Studies* 57: 41–54.

Guillot, M. 2006. Tempo effects in mortality: An appraisal. *Demographic Research* 14: 1–26.

Halli, S.S. and Rao, K.V. 1992. *Advanced Techniques of Population Analysis*. New York: Plenum.

Hanushek, E.A. and Jackson, J.E. 1977. *Statistical Methods for Social Scientists*. New York: Academic Press.

Hobcraft, J., Menken, J., and Preston, S. 1982. Age, period and cohort effects in demography: A review. *Population Index* 48: 4–43.

Hoem, J.M. and Fong, M.S. 1976. A Markov Chain Model of Working Life Tables. Working Paper No. 2, Laboratory of Actuarial Mathematics, University of Copenhagen.

Horiuchi, S. 2005. Tempo effect on age-specific death rates. *Demographic Research* 13: 189–200.

Horiuchi, S. and Preston, S.H. 1988. Age-specific growth rates: The legacy of past population dynamics. *Demography* 25: 429–41.

Horn, R.A. and Johnson, C.R. 1985. *Matrix Analysis*. Cambridge: Cambridge University Press.

Iannelli, M., Martcheva, M., and Milner, F.A. 2005. *Gender-Structured Population Modeling: Mathematical Methods, Numerics, and Simulations*. Philadelphia: Society for Industrial and Applied Mathematics.

Impagliazzo, J. 1985. *Deterministic Aspects of Mathematical Demography*. Berlin: Springer Verlag.

Jordan, C.W. Jr. 1975. *Life Contingencies* (2d ed). Chicago: Society of Actuaries.

Kannisto, V., Lauritsen, J., Thatcher, A.R., and Vaupel, J.W. 1994. Reductions in mortality at advanced ages: Several decades of evidence from 27 countries. *Population and Development Review* 20: 793–810.

Kendall, M.G. and Stuart, A. 1958. *The Advanced Theory of Statistics*. Volume I. New York: Hafner.

Keyfitz, N. 1970. Finding probabilities from observed rates, or how to make a life table. *The American Statistician* 24: 28–33.

Keyfitz, N. 1971. On the momentum of population growth. *Demography* 8: 71–80.

Keyfitz, N. 1972. Population waves. Pp. 1–38 in T.N.E. Greville (ed), *Population Dynamics*. New York: Academic Press.

Keyfitz, N. 1977. *Introduction to the Mathematics of Population* (2d ed). Reading MA: Addison-Wesley.

Keyfitz, N. 1985. *Applied Mathematical Demography*.(2d ed) New York: Springer-Verlag.

Keyfitz, N. 1988. Some demographic properties of transfer schemes: How to achieve equity between the generations. Pp. 92–105 in R.D. Lee, W.B. Arthur and G. Rogers (eds), *Economics of Changing Age Distributions in Developed Countries*. Oxford: Clarendon Press.

Keyfitz, N. and Flieger, W. 1968. *World Population: An Analysis of Vital Data*. Chicago: University of Chicago Press.

Keyfitz, N. and Flieger, W. 1971. *Population: Facts and Methods of Demography*. San Francisco: Freeman.

Keyfitz, N. and Flieger, W. 1990. *World Population Growth and Aging*. Chicago: University of Chicago Press.

Kim, Y.J. 1986. Examination of the generalized age distribution. *Demography* 23: 451–61.

Kim, Y.J. 1987. Dynamics of populations with changing rates: Generalization of the stable population theory. *Theoretical Population Biology* 31: 306–22.

Kim, Y.J. and Schoen, R. 1993a. On the intrinsic force of convergence to stability. *Mathematical Population Studies* 4: 89–102.

Kim, Y.J. and Schoen, R. 1993b. Crossovers that link populations with the same vital rates. *Mathematical Population Studies* 4: 1–19.

Kim, Y.J. and Schoen, R. 1996. Populations with quadratic exponential growth. *Mathematical Population Studies* 6: 19–33.

Kim, Y.J. and Schoen, R. 1996a. Populations with sinusoidal birth trajectories. *Theoretical Population Biology* 50: 105–23.

Kim, Y.J. and Schoen, R. 1997. Population momentum expresses population aging. *Demography* 34: 421–27.

Kim, Y.J. and Schoen, R. 1999. Changes in timing and the measurement of fertility. Paper presented at the Annual Meeting of the Population Association of America in New York.

Kim, Y.J. and Schoen, R. 2000. On the quantum and tempo of fertility: Limits to the Bongaarts-Feeney adjustment. *Population and Development Review* 26: 554–59.

Kim, Y.J., Schoen, R., and Sarma, P.S. 1991. Momentum and the growth-free segment of a population. *Demography* 28: 159–73.

Kohler, H.P. and Ortega, J.A. 2002. Tempo-adjusted period parity progression measures, fertility postponement, and completed cohort fertiltiy. *Demographic Research* 6: 91–144.

Kruithof, J. 1937. Calculation of telephone traffic. *De Ingenieur* 52: E15–E25.

Land, K.C. and Rogers, A. (eds) 1982. *Multidimensional Mathematical Demography*. New York: Academic.

Ledent, J. 1978. The dynamics of two demographic models of urbanization. Research Memorandum RM-78-56, International Institute for Applied Systems Analysis, Laxenburg, Austria.

Lee, R.D. 1974. The formal dynamics of controlled populations and the echo, the boom and the bust. *Demography* 11: 563–85.

Lee, R.D. 1980. Aiming at a moving target: Period fertility and changing reproductive goals. *Population Studies* 34: 205–26.

Lee, R.D. 1997. Population dynamics: Equilibrium, disequilibrium, and consequences of fluctuations. Pp. 1063–1115 in M.R. Rosenzweig and O. Stark (Eds), *Handbook of Population and Family Economics*, Vol 1B. Amsterdam: Elsevier.

Lee, R.D. and Carter, L.R. 1992. Modeling and forecasting U.S. mortality. *Journal of the American Statistical Association* 87: 659–72.

Leslie, P.H. 1945. On the use of matrices in certain population mathematics. *Biometrika* 33: 183–212.

Lesthaeghe, R. and Willems, P. 1999. Is low fertility a temporary phenomenon in the European Union? *Population and Development Review* 25: 211–29.

Lewis, E.G. 1942. On the generation and growth of a population. *Sankhya: The Indian Journal of Statistics* 6: 93–96.

Lexis, W. 1878. Sur la duree normale de la vie humaine et sur la theorie de la stabilite des rapports statistiques. *Annales de Demographie Internationale* 2: 447–60.

Li, N. and Lee, R. 2005. Coherent mortality forecasts for a group of populations: An extension of the Lee-Carter method. *Demography* 42: 575–94.

Li, N. and Tuljapurkar, S. 1999. Population momentum for gradual demographic transitions. *Population Studies* 53: 255–62.

Li, N. and Tuljapurkar, S. 2000. The solution of time-dependent population models. *Mathematical Population Studies* 7: 311–29.

Liaw, K-L. 1980. Multistate dynamics: The convergence of an age-by-region population system. *Environment and Planning A* 12: 589–613.

Lotka, A.J. 1907. Relation between birth rates and death rates. *Science* 26(N.S.): 21–22.

Lotka, A.J. 1939 (orig). *Analytical Theory of Biological Populations*. Translated and with an Introduction by D.P. Smith and H. Rossert. New York: Plenum Press, 1998.

Lutz, W., O'Neill, B.C., and Scherbov, S. 2003. Europe's population at a turning point. *Science* 299: 1991–92.

May, R.M. 1974. Biological populations with nonoverlapping generations: stable points, stable cycles, and chaos. *Science* 186: 645–47.

McFarland, D.D. 1975. Models of marriage formation and fertility. *Social Forces* 54: 66–83.

Menken, J. 1985. Age and fertility: How late can you wait? *Demography* 22: 469–83.

Miller, M.D. 1946. *Elements of Graduation*. Actuarial Monograph No. 1, The Actuarial Society of America and the American Institute of Actuaries.

Mitra, S. 1976. Influence of instantaneous fertility decline to replacement level on population growth: An alternative model. *Demography* 13: 513–19.

Mitra, S. 1987. Models of birth trajectories with certain patterns of variation in vital rates. *Genus* 43: 1.14.

Nair, P.S. 1985. Estimation of period specific gross migration flows from limited data: Bi-proportional adjustment approach. *Demography* 22: 133–42.

Namboodiri, N.K. 1969. On the dependence of age structure on a sequence of mortality and fertility schedules: An exposition of a cyclical model of population change. *Demography* 6: 287–99.

National Center for Health Statistics. 1997. U.S. Decennial Life Tables for 1989–91. Volume 1, Number 1. Hyattsville MD.

Ni Bhrolchain, M. 1992. Period paramount? A critique of the cohort approach to fertility. *Population and Development Review* 18: 599–629.

Noble, B. and Daniel, J.W. 1988. *Applied Linear Algebra*. Englewood Cliffs NJ: Prentice-Hall.

Omran, A. 1971. The epidemiological transition. *Milbank Memorial Fund Quarterly* 49: 509–38.

Philipov, D. 1978. Migration and settlement in Bulgaria. *Environment and Planning A* 10: 593–617.

Pollard, J.H. 1973. *Mathematical Models for the Growth of Human Populations*. Cambridge: Cambridge University Press.

Pollard, J.H. and Valkovics, E.J. 1992. The Gompertz distribution and its applications. *Genus* 48: 15–28.

Preston, S. 1975. Estimating the proportion of American marriages that end in divorce. *Sociological Methods and Research* 3: 435–60

Preston, S.H. 1978. New developments in the analysis of nuptiality and family formation and dissolution. Pp 203–18 in *Proceedings: International Population Conference, Mexico 1977*. Liege, Belgium: International Union for the Scientific Study of Population.

Preston, S.H. 1982. Relations between individual life cycles and population characteristics. *American Sociological Review* 47: 253–64.

Preston, S.H. 1983. An integrated system for demographic estimation from two age distributions. *Demography* 20: 213–26.

Preston, S.H. 1984. Children and the elderly: Divergent paths for America's dependents. *Demography* 21: 435–57.

Preston, S.H. and Coale, A.J. 1982. Age structure, growth, attrition, and accession: A new synthesis. *Population Index* 48: 217–59.

Preston, S.H., Heuveline, P. and Guillot, M. 2001. *Demography: Measuring and Modeling Population Processes*. Oxford UK: Blackwell.

Preston, S.H., Himes, C. and Eggers, M. 1989. Demographic conditions responsible for population aging. *Demography* 26: 691–704.

Preston, S.H., Keyfitz, N. and Schoen, R. 1972. *Causes of Death: Life Tables for National Populations*. New York: Seminar Press.

Preston, S.H. and McDonald, J. 1979. The incidence of divorce within cohorts of American marriages contracted since the Civil War. *Demography* 16: 1–25.

Rees, P.H. and Wilson, A.G. 1977. *Spatial Population Analysis*. London: Arnold.

Robine, J-M, Cheung, K.S.L., Tu, E. and Zeng, Y. 2003. The demographic window: Challenges and opportunities. Paper presented at the April Annual Meeting of the Population Association of America in Minneapolis.

Rodriguez, G. 2006. Demographic translation and tempo effects: An accelerated failure time perspective. *Demographic Research* 14: 85–110.

Rogers, A. 1975. *Introduction to Multiregional Mathematical Demography*. New York: Wiley.

Rogers, A. 1995. *Multiregional Demography*. Chichester, West Sussex: Wiley.

Rogers, A. and Castro, L.J. 1986. Migration. Chapter 5 in A. Rogers and F.J. Willekens (eds) *Migration and Settlement: A Multiregional Comparative Study*. Dordrecht: Reidel.

Rogers, A. and Willekens, F. 1978. The spatial reproductive value and the spatial momentum of zero population growth. *Environment and Planning A* 10: 503–18.

Rogers, A. and Willekens, F.J. 1986. *Migration and Settlement: A Multiregional Comparative Study*. Dordrecht, Holland: Reidel.

Ryder, N.B. 1964. The process of demographic translation. *Demography* 1: 74–82.

Ryder, N. 1965. The cohort as a concept in the study of social change. *American Sociological Review* 30: 854–61.

Ryder, N.B. 1969. The emergence of a modern fertility pattern: United States 1917–66. Pp. 99–126 in S.J. Behrman, L. Corsa, and R. Freedman (eds), *Fertility and Family Planning: A World View*. Ann Arbor: University of Michigan Press.

Ryder, N.B. 1975. Notes on stationary populations. *Population Index* 41: 3–28.

Ryder, N.B. 1980. Components of temporal variations in American fertility. Pp. 15–54 in R.W. Hiorns (ed), *Demographic Patterns in Developed Societies*. London: Taylor and Francis.

Ryder, N.B. 1986. Observations on the history of cohort fertility in the United States. *Population and Development Review* 12: 617–43.

Schlögl, F. 1976. Mixing distance and stability of steady states in statistical nonlinear thermodynamics. *Zeitschrift für Physik B* 25: 411–21.

Schoen, R. 1978. Calculating life tables by estimating Chiang's *a* from observed rates. *Demography* 15: 625–35.

Schoen, R. 1987. *United States Marital Status Life Tables for Periods 1910–1975 and Cohorts Born 1888–1945*. NTIS Accession No. PB87-222485/AS.

Schoen, R. 1988a. *Modeling Multigroup Populations*. New York: Plenum Press.

Schoen, R. 1988b. Practical uses of multistate population models. *Annual Review of Sociology* 14: 341–61.

Schoen, R. 2001. Toward a general model for populations with changing rates. *Demographic Research* 4: 163–84.

Schoen, R. 2002. On the impact of spatial momentum. *Demographic Research* 6: 49–62.

Schoen, R. 2003. Dynamic populations with uniform natural increase across states. *Mathematical Population Studies* 10: 195–210.

Schoen, R. 2004. Timing effects and the interpretation of period fertility. *Demography* 41: 801–19.

Schoen, R. 2005. Intrinsically dynamic population models. *Demographic Research* 12: 51–75.

Schoen, R. 2006. Insights from Parity Status Life Tables for the 20th century U.S. *Social Science Research* 35: 29–39.

Schoen, R. and Canudas-Romo, V. 2005. Changing mortality and average cohort life expectancy. *Demographic Research* 13: 117–42.

Schoen, R. and Canudas-Romo, V. 2005a. Timing effects on first marriage: Twentieth-century experience in England and Wales and the USA. *Population Studies* 59: 135–46.

Schoen, R. and Canudas-Romo, V. 2005b. Multistate cohort analysis with proportional transfer rates: An application to marriage, divorce, and remarriage. Paper presented at the July General Conference of the International Union for the Scientific Study of Population, Tours, France.

Schoen, R. and Canudas-Romo, V. 2006. Multistate cohort models with proportional transfer rates. *Demography*, 43: 553–568.

Schoen, R. and Canudas-Romo, V. 2006a. Timing effects on divorce: Twentieth century experience in the United States. *Journal of Marriage and Family*, 68: 749–758.

Schoen, R., Greenblatt, H.N. and Mielke, R.B. 1975. California's experience with non-adversary divorce. *Demography* 12: 223–43.

Schoen, R. and Jonsson, S.H. 2003. Modeling momentum in gradual demographic transitions. *Demography* 40: 621–35.

Schoen, R. and Jonsson, S.H. 2003a. A diminishing population whose every cohort more than replaces itself. *Demographic Research* 9: 111–18.

Schoen, R. and Jonsson, S.H. 2003b. Estimating multistate transition rates from population distributions. *Demographic Research* 9: 1–24.

Schoen, R. and Jonsson, S.H. 2006. Some intergenerational transfer implications of birth fluctuations. Pp. 279–290 in A.H. Gauthier, C.Y.C. Chu, and S. Tuljapurkar (eds). *Allocating Public and Private Resources Across Generations*. New York: Springer.

Schoen, R., Jonsson, S.H., and Tufis, P. 2004. A population with continually declining mortality. Working Paper 04–07, Population Research Institute, Pennsylvania State University.

Schoen, R. and Kim, Y.J. 1991. Movement toward stability as a fundamental principle of population dynamics. *Demography* 28: 455–66.

Schoen, R. and Kim, Y.J. 1992. Covariances, roots, and the dynamics of age-specific growth. *Population Index* 58: 4–17.

Schoen, R. and Kim, Y.J. 1993. Two-state spatial dynamics in the absence of age. *Theoretical Population Biology* 44: 67–79.

Schoen, R. and Kim, Y.J. 1994. Cyclically stable populations. *Mathematical Population Studies* 4: 283–95.

Schoen, R. and Kim, Y.J. 1994a. Hyperstability. Paper presented at the Annual Meeting of the Population Association of America, Miami, May 5–7.

Schoen, R. and Kim, Y.J. 1996. Stabilization, birth waves, and the surge in the elderly. *Mathematical Population Studies* 6: 35–53.

Schoen, R. and Kim, Y.J. 1997. Exploring cyclic net reproduction. *Mathematical Population Studies* 6: 277–90.

Schoen, R. and Kim, Y.J. 1998. Momentum under a gradual approach to zero growth. *Population Studies* 52: 295–99.

Schoen, R. and Kim, Y.J. 2000. A dynamic multistate model of robustness and frailty. *Mathematical Population Studies* 8: 293–304.

Schoen, R. and Land, K.C. 1979. A general algorithm for estimating a Markov-generated increment-decrement life table with applications to marital-status patterns. *Journal of the American Statistical Association* 74: 761–76.

Schoen, R. and Standish, N. 2001. The retrenchment of marriage: Results from marital status life tables for the United States, 1995. *Population and Development Review* 27: 553–63.

Schoen, R., Urton, W., Woodrow, K., and Baj, J. 1985. Marriage and divorce in twentieth century American cohorts. *Demography* 22: 101–14.

Seneta, E. 1981. *Non-negative Matrices and Markov Chains* (2nd ed). New York: Springer-Verlag.

Sharpe, F.R. and Lotka, A.J. 1911. A problem in age-distribution. *Philosophical Magazine*, Ser. 6, 21: 435–38.

Siler, W. 1979. A competing risk model for animal mortality. *Ecology* 60: 750–57.

Skellam, J.G. 1967. Seasonal periodicity in theoretical population ecology. Pp. 179–205 in *Proceedings*, Fifth Berkeley Symposium on Mathematical Statistics and Probability, Vol IV "Biology and Problems of Health" edited by L. Lecam and J. Neyman. Berkeley: University of California.

Strogatz, S.H. 1994. *Nonlinear Dynamics and Chaos*. Westview Press.

Sykes, Z. 1973. Intrinsic age-specific birth rates: A new method of fertility analysis. Unpublished manuscript, Department of Population Dynamics, Johns Hopkins University, Baltimore.

Tiemeyer, P. and Ulmer, G. 1991. *MSLT: A Program for the Computation of Multistate Life Tables*. Center for Demography and Ecology Working Paper 91–34. University of Wisconsin-Madison

Tuljapurkar, S.D. 1982. Why use population entropy? It determines the rate of convergence. *Journal of Mathematical Biology* 13: 325–37.

Tuljapurkar, S.D. 1985. Population dynamics in variable environments. VI. Cyclical environments. *Theoretical Population Biology* 28: 1–17.

Tuljapurkar, S.D. 1990. *Population Dynamics in Variable Environments (Vol 85)*. New York: Springer.

Tuljapurkar, S.D. 1993. Entropy and convergence in dynamics and demography. *Journal of Mathematical Biology* 31: 253–71.

United Nations, Population Division. 1999. *Population Growth, Structure and Distribution: The Concise Report*. New York: United Nations.

United Nations, Population Division. 2005. *World Population Prospects: The 2004 Revision*. Vol. 1: Comprehensive Tables (ST/ESA/SER.A/244). New York: United Nations.

Van Imhoff, E. and Keilman, N. 2000. On the quantum and tempo of fertility: Comment. *Population and Development Review* 26: 549–53.

Vaupel, J.W. 1986. How change in age-specific mortality affects life expectancy. *Population Studies* 40: 147–57.

Vaupel, J.W. and Canudas-Romo, V. 2003. Decomposing change in life expectancy: A bouquet of formulas in honor of Nathan Keyfitz's 90th birthday. *Demography* 40: 201–16.

Vogelsang, H. and Feichtinger, G. 1979. Exponentielle Niveau-und musteranderung der fertilitat. *Zeitschrift fur Bevolkerungswissenschaft* 1: 31–63.

Von Förster, H. 1959. Some remarks on changing populations. In F. Stohlman, Jr. (ed) *The Kinetics of Cellular Proliferation*. New York: Greene and Stratton.

Wachter, K.W. 1988. Age group growth rates and population momentum. *Population Studies* 42: 487–94.

Wachter, K.W. 1991. Elusive cycles: Are there dynamically possible Lee-Easterlin models for U.S. births? *Population Studies* 45: 109–35.

Wachter, K.W. 1994. The cohort feedback model with symmetric net maternity. *Mathematical Population Studies* 5: 25–44.

Wachter, K.W. 2005. Tempo and its tribulations. *Demographic Research* 13: 201–22.

Wachter, K.W. and Lee, R.D. 1989. U.S. births and limit cycle models. *Demography* 26: 99–115.

Willekens, F. 1982. Multidimensional population analysis with incomplete data. Pp. 43–111 in K.C. Land and A. Rogers (eds), *Multidimensional Mathematical Demography*. New York: Academic Press.

Willekens, F. 1999. Modeling approaches to the indirect estimation of migration flows: From entropy to E.M. *Mathematical Population Studies* 7: 239–78.

Wilmoth, J.R. 1997. In search of limits. Pp. 38–64 in K.W. Wachter and C.E. Finch (eds), *Between Zeus and the Salmon: The Biodemography of Longevity*. Washington DC: National Academy Press.

Wilmoth, J.R. and Horiuchi, S. 1999. Rectangularization revisited: Variability of age at death within human populations. *Demography* 36: 475–95.

Winkler-Dworak, M. and Engelhardt, H. 2004. On the tempo and quantum of first marriages in Austria, Germany, and Switzerland: Changes in mean and variance. *Demographic Research* 10: 231–64.

Wright, G.C. and Stetson, D.M. 1978. The impact of no-fault divorce law reform on divorce in American states. *Journal of Marriage and the Family* 40: 575–80.

Zeng, Y. and Land, K.C. 2002. Adjusting period tempo changes with an extension of Ryder's basic translation equation. *Demography* 39: 269–85.

# AUTHOR INDEX


**A**

Alho, J.M., 82
Alligood, K.T., 124
Arthur, W.B., 42, 61, 62, 64

**B**

Baj, J., 119
Batty, M., 189
Bennett, N.G., 63
Bernardelli, H., 12
Birdsall, N., 37
Bishop, Y.M.M., 189
Bongaarts, J., 44, 81, 82, 85–89, 101, 103, 104, 112
Bouvier, L.F., 42
Boyce, W.E., 177, 182
Brouard, N., 82
Bulatao, R.A., 44, 81
Butz, W.P., 105, 106, 110

**C**

Calot, G., 104
Canudas-Romo, V., xi, 83, 84, 96, 97, 99, 100, 114, 115, 118, 119, 162, 164, 165, 167, 169–173
Carter, L.R., 82
Castro, L.J., 55
Caswell, H., 9, 27, 29, 217
Cerone, P., 50
Cheung, K.S.L., 37
Chilton, R., 190
Chiang, C.L., 5, 14
Coale, A.J., 45, 48, 54, 62–64, 69, 76, 80, 81, 107, 123, 132, 133
Cohen, J.C., 183

**D**

Daniel, J.W., 35
Demeny, P., 45, 48, 80, 81
Deming, W.E., 181
DeMoivre, A., 23
DeSantis, G., 131
DiPrima, R.C., 177, 182
Dupasquier, L.G., 14

**E**

Easterlin, R.A., 128
Eggers, M., 69, 76, 79
Elo, I.T., 83
Engelhardt, H., 113
Espenshade, T.J., 42
Euler, L., 9

**F**

Feeney, G.M., 54, 85, 87–89, 101, 103, 104, 112, 203
Feichtinger, G., 132
Feller, W., 9
Fienberg, S.E., 189
Fisher, R.A., 30
Fix, E., 14
Flieger, W., 32, 32, 48
Fong, M.S., 19
Ford, L.R., 39, 175, 181
Franklin, J.N., 217
Frauenthal, J.C., 50, 124
Frejka, T., 104
Furness, K.P., 189

**G**

Gantmacher, F.R., 143, 183, 217
Goldstein, J.R., 50, 51, 89, 94–96, 114
Graunt, J., 1, 79
Greenblatt, H.N., 118
Guillot, M., 2,6,8,77,82, 89, 99, 193

**H**

Halli, S.S., 189
Hanushek, E.A., 69

Heuveline, P., 2, 6, 8, 77, 193
Himes, C., 69, 76, 79
Hobcraft, J., 83
Hoem, J.M., 19
Holland, P.W., 189
Horiuchi, S., 63, 64, 96, 122
Horn, R.A., 217

**I**

Iannelli, M., xi
Impagliazzo, J., 35

**J**

Jackson, J.E., 69
Johnson, C.R., 217
Jonsson, S.H., xi, 51, 82, 93, 99, 112, 113, 117, 129–132, 138, 140, 141, 188, 190, 192–195
Jordan, C.W., Jr, 23

**K**

Kannisto, V., 79
Keilman, N., 104
Kendall, M.G., 66
Keyfitz, N., 4, 6, 9, 12, 31–33, 43, 44, 48, 60, 66, 72, 76, 129, 149
Kim, Y.J., xi, 29, 31–33, 35, 36, 39, 40, 45–50, 52–54, 64–67, 69–74, 76, 104, 107, 121, 123–126, 128, 132, 133, 136–138, 153–156, 181, 222
Kohler, H.-P., 104
Kruithof, J., 189

**L**

Land, K.C., 4, 14, 104
Lauritsen, J., 79
Ledent, J., 39
Lee, R.D., 82, 102, 123, 124, 128
Leslie, P.H., 12
Lesthaeghe, R., 104
Lewis, E.G., 12
Lexis, W., 96
Li, N., 50, 82
Liaw, K.-L., 40
Lotka, A.J., 9–11
Lutz, W., 44, 105

**M**

Mackie, S., 189
Malthus, T.R., 128
Martcheva, M., xi
May, R.M., 124
McDonald, J., 116
McFarland, D.D., 190
Menken, J., 83, 129
Mielke, R.B., 118
Miller, M.D., 6, 24
Milner, F.A., xi
Mitra, S., 50, 133

**N**

Nair, P.S., 190
Namboodiri, N.K., 123
National Center for Health Statistics (U.S.), 207
Neyman, J., 14
NiBhrolchain, M., 102, 110
Noble, B., 35

**O**

Omran, A., 79
O'Neill, B.C., 44, 105
Ortega, J.A., 104

**P**

Perron and Frobenius, 28
Philipov, D., 190
Poet, R., 190
Pollard, J.H., 9, 94
Preston, S.H., 2, 4, 6, 8, 46, 62–66, 69, 76, 77, 79, 83, 116, 129, 193

**R**

Rao, K.V., 189
Rees, P.H., 19
Robine, J.-M., 37
Rodriguez, G., 89

Rogers, A., 14, 20, 51, 55, 161
Ryder, N.B., 3, 50, 52, 81, 83, 102, 105, 108, 110, 121

**S**

Sarma, P.S., 46, 47, 49
Sauer, T.D., 124
Scherbov, S., 44, 105
Schlögl, F., 29
Schoen, R., xi, 2, 4, 5, 8–10, 14, 20, 21, 29, 31–33, 35–37, 39, 40, 45–56, 65–67, 69–74, 76, 82–84, 93, 96, 99, 101, 104, 106–109, 111–115, 118, 119, 121, 124–126, 128–133, 136–138, 140–143, 147, 149–153, 155, 156, 162, 164, 165, 167, 169–174, 177, 180, 181, 183, 187, 188, 190, 192–195, 222
Seneta, E., 217
Sharpe, F.R., 9
Siler, W., 85, 100
Sinding, S.W., 37
Skellam, J.G., 123
Spencer, B.D., 82
Standish, N., 118, 194
Stecklov, G., 51
Stephan, F.E., 189
Stetson, D.M., 118, 119
Strogatz, S.H., 124
Stuart, A., 66
Swick, K.E., 124
Sykes, Z., 73, 133

**T**

Thatcher, A.R., 79
Tiemeyer, P., 20
Trussell, T.J., 54, 107, 133
Tu, E., 37
Tufis, P., 82, 93, 99
Tuljapurkar, S.D., 29, 32, 50, 123, 124, 143, 148

**U**

Ulmer, G., 20
United Nations, Population Division, 55, 203
Urton, W., 119

**V**

Valkovics, E.J., 94
Van Imhoff, E., 104
Vaupel, J.W., 64, 79, 92, 93, 99
Vogelsang, H., 132
Von Förster, H., 64

**W**

Wachter, K.W., 46, 89, 94, 95, 124, 128
Ward, M.P., 105, 106, 110
Willekens, F.J., 51, 161, 189, 190
Willems, P., 104
Wilmoth, J.R., 81, 96
Wilson, A.G., 19
Winkler-Dworak, M., 113
Woodrow, K., 119
Wright, G.C., 118

**Y**

Yorke, J.A., 124

**Z**

Zeng, Y., 37, 104

# SUBJECT INDEX


**A**

Absorbing ("dead") state, 14
Adjusted First Marriage Ratio (AFMR), 115
Age
  at death
    modal age, 81, 96–98
    relationship to CAL, 88
    standard deviation of, 81, 99 (#1)
  intervals, defined, 2
  mean ages
    of childbearing, 11, 39, 44, 46, 76 (#5), 144–146, 148, 150, 152, 157 (#2), 178–179
    of population, 11, 47, 60, 69, 76 (#6), 80–81
    of reproductivity, 47
    of stable net maternity, 137
  specific rates, 2
Age-Period-Cohort effects in MASH model, 216 (#6)
Age dependent growth, 63–65, 136–137
  characteristic equation, 66, 76 (#4)
  differentiating demographic measures, 67–70
  in shifting Gompertz model, 88
  in metastable model, 136–137
  in spectral decomposition, 73–74
  relation to intrinsic r, 71
Aging, see Population Aging
Autonomous growth rate, 38
Average Cohort Fertility (ACF), 105–110
Average Cohort Life Expectancy (ACLE), 83–85

**B**

Bongaarts-Feeney model
  of fertility, 103–105, 106–110, 112–113
  of mortality, 85–89
  mean age at childbearing (MAC), 103

**C**

CAL, 82–84, 87–89, 91, 92–96, 207–208
Change, marginal, see Marginal change
Chiang's a, 5
Coale-Demeny Model "West"
  life tables, 45
  mortality change in, 80–81
  stable populations, 48
Coefficient of Variation, 32
Cohort
  defined, 3
  gaps and lags in life expectancies, 94–96
  fertility, period-cohort contrast, 101–103, 112–113
  mortality, period-cohort contrast, 82–85
  stable population, 11, 62–63
Component, of population, see Leslie matrix, spectral decomposition
Constant forces assumption
  basic life table, 3, 5
  multistate, 18
Constant generation length models, 124–128, 149
  cyclical models, 126–128, 149
Constant proportional distribution of births by age, 136, 137, 153
Convergence to stability
  at the margin, 70–72
  conditions, 27–29
  empirical measures, 33
  entropy, 32–33
  ergodicity, 28
  force of convergence, 31–32, 39–40
  fundamental principle, 72
  hyperstability, 154
  in 2x2 Leslie matrices, 34
  in 2 living state models, 37–40
  in 3x3 Leslie matrices, 34–37
  Kullback distance, 29–31, 39
  metastability, 134
  primitivity, 28–29
  role of eigenvalues, 27–29
  spatial momentum calculations, 52–56
  speed, 29–33, 34, 39–40
  trajectories, 36

Covariances, 69–70
Crossover points, 65–67
  relation to Lotka's r, 66–67
  relation to Lotka's Ŧ, 66
  transformation of summary measures, 67
Cubic equation, solution, 158–159 (#13)
Cyclically stable/stationary models, 123–124, 129–131, 155, 176–178
Cumulants, 66

**D**

Demography, 1, 214
Dependency burden, 37, 129–130
Derivatives of demographic measures, 67–70
Differential equation, basic demographic, 213–214
Double root (in 3x3 Leslie matrix), 35, 42 (#5), 58 (#3)
Dynamic modeling, definition, xi

**E**

Easterlin Hypothesis, 128, 156–157 (#1)
Eigenstructure, 27–29, 218–220
Entropy
  of net maternity, 32
  of survivorship, 60, 92–93
  relation to force of convergence, 32–33
  relation to IPF, 189
Ergodicity, 28
Euler's constant, 94
Expectancies, see Life expectancies
Expectation of Productive Life, 23
Exponential approach, see constant forces assumption

**F**

Fertility, 9–10, 12, 21
  models with proportional rates, 155
  timing of, 101–113
    Average Cohort Fertility (ACF), 105–110
    Bongaarts-Feeney approach, 103–105, 106–110
  transitions, 138–142
Fixed f assumption, see Hyperstable models
Flow equations, see General Algorithm
Forces (intensities)
  cause-specific, 8
  convergence, 31, 39–40, 53
  constant forces approach, 3, 5
  force of mortality
    definition, 2
    marginal change in, 59–60
  IDM multistate, 174–175
  multistate, 15
Functional differentiation, 61–62

**G**

Gaps and lags in life expectancies, 94–96
General Algorithm (for life table construction)
  basic life table, 4
  flow equations, 4, 14, 192
  orientation equations, 4
  person-year equations, 4–5
  multistate life table, 14–17
  multistate stable population, 22
  renewal equations, 10, 21
Gompertz Law, 23 (#5), 84
Graduation (smoothing) 6, 24 (#12)
Gross flows, 14–15
  equations, 15

**H**

Hadwiger function, 31–32, 33
Hyperstable models
  birth-death, 153–154, 159 (#15)
  fixed fraction of births by age, 136, 137, 153
  MASH model, 204–208
  mortality model, 159 (#15)
  multistate, 181–185
    basic differential equations, 181–182, 183–184
    Additive Eigenvector assumption, 183, 188, 200 (#11)
    latent root matrix ($\boldsymbol{\nu}$), 183
    population projection matrix, 182, 184
    product matrix, 182–183
    proportional distribution by origin, 184
    Proportional eigenvector assumption, 183, 186–188
    two living state model, 185–188
  relationships between multi-age and multistate models, 184–185

I

Increment-decrement life tables, see Multistate life tables
Indirect standardization, 193
Integrals, implicit range, 9, 44
Intergenerational relationships with cyclical births, 128–131
Intrinsically Dynamic Model (IDM), 143–152, 174–181
  birth-death model, 143–152
  birth trajectories, 149
  cyclical models, 176–178, 199 (#8)
  differential equation specification, 174–175, 180, 181
  momentum, 150–152
  multiplicative property, 143, 144
  multistate model, 174–181
  n age group IDM, 148–149, 181
  relationship to observed populations, 150
  three age group IDM, 146–148, 180
  two age group IDM, 143–145, 174–179, 199 (#7), 199 (#8)
  uniform natural increase (growth) assumption, 174
Intrinsic birth rate, 11, 43
Intrinsic growth rate, 10, 12, 21, 44, 71
Iterative Proportional Fitting (IPF), 189–190, 193–196
  description of method, 189–190
  comparison with RSA, 193–196
Iterative solutions
  stable population r, 12
  multistate stable population r, 21

J

Jordan matrix, 35, 42 (#7)

K

Kullback Distance, 29–31, 39, 71–72, 138, 174, 198 (#6), 199 (#7)

L

Lags, see Gaps and lags in life expectancies
Length of generation (Lotka's Ŧ), 11, 46, 65, 66, 125
Leslie matrix
  block matrix, 53–54, 204, 210
  defined, 12
  eigenstructure, 13
  elements in terms of roots, 143, 146, 148
  metastable form, 132
  spectral decomposition, 72–74
  3x3 case, 34–37, 58 (#3), 58 (#4), 146–148
    solution for eigenvalues, 158–159 (#13)
  2x2 case, 34, 143–145
Lexis surface, 82
Life expectancies
  at birth, 43
  basic life table, 4
  population based, 19, 164
  status based, 20
Life table
  ASDT, 7–8
  basic, 3–7
  cause eliminated, 7–8
  construction
    constant forces approach, 3, 4–5, 18
    early and late ages, 6–7, 18
    linear assumption, 5, 18
    mean duration at transfer approach, 5, 18
  early efforts, 1
  multiple decrement, 7–8
  multistate, 14–20
  parity status, 111
  stationary population, 6
Linear assumption
  basic life table, 5
  multistate life tables, 18
Linearly changing mortality, 92–94
  extended Gompertz linear change model, 94
  Gompertz linear change model, 92–93, 96, 97–98
  linearly shifting Gompertz model, 93–94, 95, 96
Lotka's r, 10, 218 (see also Intrinsic growth rate)

M

Marginal change
  convergence to stability, 70–72
  covariance relationships, 69–70
  derivatives of demographic measures, 67–70, 76 (#9)

functional differentiation, 61–62
in age and time, 63–64
in metastable models, 138–142
in mortality, 59–60
in stable population characteristics, 62–63
on age composition, 60–62, 68
on cohort characteristics, 62–63
on intrinsic growth, 61
on period characteristics, 62–63, 68–70
spectral decomposition of, 72–74
Markov
assumption, 15
transition probabilities, 15
Mathematical software, 20, 192, 219
Matrix(es)
basic algebra, 217–220
eigenstructure, 27–29, 218–220
imprimitive, 28, 41 (#3)
multistate life table equations, 16–18
primitive, 28
rank one, 28, 224
reducible, 28–29, 41 (#4)
spectral decomposition, 72
Mean ages, see under Age
Mean duration at transfer approach
basic life table, 5
multistate life tables, 18
Mean Value Theorem, 125
Metabolism, 175
Metastable models
age composition, 134, 136
associated stable population, 137
birth-death models, 132–142
birth trajectory, 135
continuous case, 135–138
discrete case, 132–135
fixed fraction of births by age, 136, 137, 153
Gaussian birth trajectory, 157 (#4)
growth rate, 135, 136–137
Kullback distance, 138, 198 (#6)
MASM model, 209–211
momentum, 137, 138–142, 174, 198 (#6)
multistate, 172–174, 185–188, 198 (#6)
two living state model, 185–188
product matrix, 134, 173
relation to hyperstable models, 153, 158 (#12)
Modal age at death, 96–98
Momentum, see Population momentum
Mortality (see also Mortality, force of)
cohort effects, 83
hyperstable model of changing, 159 (#15)
linearly changing mortality models, 92–94
extended Gompertz linear change model, 94
Gompertz linear change model, 92–93, 96, 97–98
linearly shifting Gompertz model, 93–94, 95, 96
overview, 79–82, 96, 159
shifting Gompertz change model, 85–91
Bongaarts-Feeney approach, 85–89
further relationships, 89–91
Mortality, force of (also see Mortality; Forces, force of mortality)
at modal age, 97
definition, 2
Gompertz Law, 23 (#5), 84
multistate, 15
neutral change, 60, 62–63
shape of curve, 80
Siler model, 85, 100 (#6)
Moving average, 24 (#12)
Multi-Age and State Hyperstable (MASH) model, 204–208
age-state composition, 206
MASH/MASM model, 215–216 (#5)
population projection matrix, 204–206
product matrix, 206
Robust/Frail model application, 207–208
Multi-Age and State Metastable (MASM) model, 209–211
age-state composition, 210
MASH/MASM model, 215–216 (#5)
population projection matrix, 209–211
product matrix, 211
renewal equation, 210–211, 214 (#2), 214–215 (#4)
Multi-age dynamic models
birth-death, 123–124, 128–159
urban-rural (multistate), 52–56
relationship to multistate, 184–185
with multiple living states, 203–206
Multistate models
computer programs, 20
convergence, 34–37, 206, 211
expectancies, 19–20

force of convergence, 39–40
gross flows, 14–15
hyperstable, 181–185
life tables, 14–20
matrix equations, 16–18
metastable, 172–174
multi-age, 52–56, 203–216
net flows, 14–15
Proportional Transfer Rate (PTR) models, 161–172, 173, 186
prospective and retrospective data, 19
rates consistent with population distributions, 188–196
spatial momentum, 51–56
stable populations, 20–22
survivorship proportions, 19
two living state
eigenstructure, 38
population projection matrix, 197 (#1)
solution to equations, 38–9

**N**

Natural increase, rate of, 52, 93
Net flows, 14
equations, 15
Net maternity function
cumulants, 66
definition, 10, 124
entropy, 32
moments, 10, 32, 47
parameterizations, 31–32
stable, 31–32, 121 (#2)
Net Reproduction Rate (NRR), 10, 44, 124, 144, 146, 148
Nonlinear models, 124
Normal curve as a metastable trajectory, 157 (#4)

**O**

Occurrence/exposure rates
age-specific, 2
definition, 1
Orientation equations, see General Algorithm

**P**

Parity, 104, 110–111, 166–170
parity-specific PTFR, 110–111
PTR models, 167–170
Period (in demography)
defined, 3
fertility, period-cohort contrast, 101–103, 112–113
gaps and lags in life expectancies, 94–96
marginal changes in measures, 67–70
mortality, period-cohort contrast, 82–85
rates with additional specificity, 110–111
stable population, 11, 62–63
Person-years
defined, 4
equations, see General Algorithm
in PTR model, 164
Phase difference, in cyclical model, 127, 176–177
Population aging
demographic "window" or "bonus", 37
momentum related, 46–50
"surge", 37
Population momentum, 43–58
age-time-specific, 45, 198 (#6)
aging effect, 46–50
basic dynamics, 43–45
defined, 43
discrete form, 73–74
distance from stability, 71–72
fundamental principle of dynamics, 72
Goldstein momentum, 50–51
Goldstein-Stecklov momentum, 51
gradual declines to stationarity, 50–51, 138–142, 150–152
growth associated with, 44, 48–49, 66–67
Intrinsically Dynamic Models, 150–152
Keyfitz momentum, 43–44, 50–51
Li-Tuljapurkar momentum, 50
metastable, 137, 138–142
negative momentum, 45
reproductive contributions, 74
sex-composition momentum, 51
spatial momentum, 51–56
contribution to growth, 50
decomposing, 54–56
definition, 52
dynamics, 52
interaction term, 54, 55
transformations to stationarity, 52, 54
urban-rural models, 52–56
spectral decomposition of, 72–74
wedge-period population, 82

Population projection matrix (PPM)
  defined, 13
Postponement of births, see Timing effects, on fertility
Power series equation, 57
Probabilities
  Markov transition, 15
  p-probabilities, 14–15
  $\pi$-probabilities, 15
Probability a Marriage Ends in Divorce (PMED), 118
Product matrix, 133, 144, 154, 163, 173, 179, 187, 198 (#5), 206, 211
Proportion Ever Marrying (PEM), 114
Proportional Transfer Rate (PTR) models, 161–172, 173, 197 (#3)
  age-state composition, 163–164
  base rate matrix, 162
    sensitivity analyses, 168–170
  eigenstructure relationships, 163
  hierarchical models, 166–170
  life course schematic, 165, 168, 171–173
  marital status models, 170–172, 173
  MASH model, 204–206
  mortality options, 164
  parity status, 167–170
  person-years, 164
  proportionalized (stylized) rates, 167, 171–172
  special case of equal transfer rates, 197 (#3)
  two living states, 165–166

Q

Quadratic Hyperstable (QH) model, see Metastable models, birth-death
Quantum of fertility, 102

R

Radix (of life table), 3
Rates, incidence, 103
Rates, occurrence/exposure
  age-specific, 2
  consistent with multistate distributions, 189–196
    IPF method, 189–190
    RSA method, 190–192
  definition, 1
Relative amplification, in cyclical model, 127–128, 176–177, 199 (#8)
Relative State Attraction (RSA) method, 190–196, 200 (#12), 214 (#1)
  description, 190–192
  comparison with IPF, 193–196
Renewal equations, 10, 21, 136, 210–211
Reproductive
  contributions and population momentum, 74, 219
  function, 30, 70–71
  value, 30, 137
Row stochastic
  population projection matrix, 133, 201 (#15), 204
  rate matrix, 17

S

Scalar, defined, 217
Shifting Gompertz mortality change model, 85–91
  Bongaarts-Feeney approach, 85–89
  derivatives of period functions, 100 (#9)
  deriving the shift in population, 90–91
  further relationships, 89–91
  linear relationship between offsets, 89
Siler model, 85, 100
Spatial momentum, see under Population momentum
Stable equivalent, 30–31, 36, 37, 39, 43–45, 134, 137, 150–151, 198 (#6), 219
Stable population, 9–13, 123, 142, 150, 154
  age composition, 11
  changes in characteristics, 62–63
  convergence to, 27–29
  continuous models, 9–11
  discrete models, 12–13
  dynamic models, 123–124
  intrinsic birth rate, b, 11, 43
  intrinsic growth rate, r, 10, 12, 21, 44
  mean ages, 11
  minimum crude death rate, 62
  multistate, 20–22
Stationary population
  basic life table, 6

Survivorship (see also Life table)
  average cohort life expectancy, 83–85
  in basic life table, 2–3
  in multistate life table, 14–15
  in stable population, 9, 11
  marginal change in, 59–60
  modal age at death as inflection point, 96
  wedge-period perspective, 82
Sykes transformation, 54, 58 (#4), 107, 133, 201 (#15)

T

Timing effects
  defined, 105
  on divorce, 116–120
  on fertility, 103–113
    in models, 107
    in the U.S., 1917–97, 107–110
  on marriage, 113–116
  parity effects, 111
  period-cohort paradox, 112–113
Timing Index (TI), 105, 115, 118
Total Fertility Rate (TFR), 101–103
  Bongaarts-Feeney adjusted TFR, 104
  cohort TFR (CTFR), 105
  parity-specific TFR, 110–111
Total First Nuptiality Ratio (TFNR), 114
Trajectories
  of births, 135, 149, 150–151, 152, 153–154
  to stability, 36
Translation, demographic, 121
Triangular matrix, lower, 167
Two-sex population models, xi, 190

U

Urban-rural multistate model, 40, 52–56

V

Variable r, see Age dependent growth
Vitality assumption, 8

W

Wedge-period, 82, 207–208, 211
Wicksell function, 31

# Springer Series on Demographic Methods and Population Analysis

1. Robert Schoen: *Modeling Multigroup Populations*. 1987. ISBN 0-306-42649-8
2. David P. Smith: *Formal Demography*. 1992. ISBN 0-306-43869-0
3. Louis G. Pol and Richard K. Thomas: *The Demography of Health and Health Care*. 1992. ISBN 0-306-43981-6
4. S.S. Halli and K.V. Rao: *Advanced Techniques of Population Analysis*. 1992. ISBN 0-306-43997-2
5. Dudley L. Poston Jr. and David Yaukey (Eds): *The Population of Modern China*. 1992. ISBN Hb: 0-306-44235-3; Pb: 0-306-44138-1
6. Susan M. de Vos: *Household Composition in Latin America*. 1995. ISBN 0-306-44962-5
7. Oscar Harkavy: *Curbing Population Growth*: An insider's perspective on the population movement. 1995. ISBN 0-306-45050-X
8. Evert van Imhoff, Anton Kuijsten, Pieter Hooimeijer and Leo van Wissen (Eds): *Household Demography and Household Modeling*. 1995. ISBN 0-306-45187-5
9. Krishnan Namboodiri: *A Primer of Population Dynamics*. 1996. ISBN 0-306-45338-X
10. Michael Micklin and Dudley L. Poston Jr. (Eds): *Continuities in Sociological Human Ecology*. 1998. ISBN 0-306-45610-9
11. Alfred J. Lotka: *Analytical Theory of Biological Populations*. 1988. ISBN 0-306-45927-2
12. Leo J.G. van Wissen and Pearl A. Dykstra (Eds): *Population Issues*. An interdisciplinary focus. 1999. ISBN 0-306-46196-X
13. Louis G. Pol and Richard K. Thomas: *The Demography of Health and Health Care*. Second Edition. 2000. ISBN Hb: 0-306-46336-9; Pb: 0-306-46337-7
14. Stanley K. Smith, Jeff Tayman and David A. Swanson: *State and Local Population Projections*. Methodology and analysis. 2001. ISBN Hb: 0-306-46492-6; Pb: 0-306-46493-4
15. Samuel H. Preston, Irma T. Elo, Mark E. Hill and Ira Rosenwaike: *The Demography of African Americans, 1930–1990*. 2003. ISBN 1-4020-1550-X
16. William A. Kandel and David L. Brown (Eds): *Population Change and Rural Society*. 2006. ISBN Hb: 1-4020-3911-5; Pb: 1-4020-3901-8
17. Robert Schoen: *Dynamic Population Models*. 2006. ISBN 1-4020-5229-4

FSC
www.fsc.org
MIX
Papier | Fördert gute Waldnutzung
FSC® C083411

Kolibri 360 GmbH
Ferdinand-Jühlke-Straße 7
99095 Erfurt, Deutschland
produktsicherheit@kolibri360.de